HISTORY OF HINDU MATHEMATICS

A SOURCE BOOKS

BY

BIBHUTIBHUSHAN DATTA

and

AVADHESH NARAYAN SINGH

IN TWO PARTS

PART 1

COSMO PUBLICATIONS

2009 New Delhi

HISTORY OF HINDU MATHEMATICS

A SOURCE BOOK

PART I

NUMERAL NOTATION AND ARITHMETIC

HISTORY OF HINDU MATHEMATICS

A SOURCE BOOK

PART I

NUMERAL NOTATION AND ARITHMETIC

BY

BIBHUTIBHUSAN DATTA

AND

AVADHESH NARAYAN SINGH

इदं नम ऋषिभ्यः पूर्वजेभ्यः पूर्वेभ्यः पथिकृद्भ्यः

(*RV*, x. 14. 25)

To the Seers, our Ancestors, the first Path-makers

PREFACE

Little is known at present to historians of mathematics regarding the achievements of the early Hindu mathematicians and our indebtedness to them. Though it is now generally admitted that the decimal place-value system of numeral notation was invented and first used by the Hindus, it is not yet fully realized to what extent we are indebted to them for our elementary mathematics. This is due to the lack of a reliable and authentic history of Hindu mathematics. Our object in writing the present book has been to make up for this deficiency by giving a comprehensive account of the growth and development of the science of mathematics in India from the earliest known times down to the seventeenth century of the Christian era.

The subject is treated by topics. Under each topic are collected together and set forth in chronological order translations of relevant Sanskrit texts as found in the Hindu works. The texts have been elucidated, wherever necessary, by adding explanatory notes and comments, and also by illustrative examples culled from original sources. We have tried to avoid repetition as far as has been consistent with our aim. However, on several occasions it has been considered desirable to repeat the same rule in the words of different authors in order to emphasize the continuity or rather the gradual evolution of mathematical thought and terminology in India. Comparative study of this kind has helped us to throw light on certain obscure Sanskrit passages and technical terms whose full significance had not been understood before. In translating the texts we have tried to be as literal and faithful as possible without sacrificing the spirit of the original. Sometimes it has not been possible to find exact parallels to Sanskrit words and technical terms in English. In all such cases we have tried to maintain the spirit of the original in the English version.

The above plan of the book has been adopted in pursuance of our intention to place before those who have no access to the Sanskrit sources all evidence, unfavourable as well as favourable, so that they can judge for themselves the claims of Hindu mathematics, without depending solely on our statements. In order to facilitate comparison with the development of mathematics in other countries the various topics have been arranged generally in accordance with the sequence in Professor D. E. Smith's *History of Mathematics*, Vol. II. This has sometimes necessitated divergence from the arrangement of topics as found in the Hindu works on mathematics.

In search of material for the book we had to examine the literature of the Hindus, non-mathematical as well as mathematical, whether in Sanskrit or in Prâkrit (Pâlî and Ardha Mâgadhî). Very few of the Hindu treatises on mathematics have been printed so far, and even these are not generally known. The manuscript works that exist in the various Sanskrit libraries in India and Europe are still less known. We have not spared labour in collecting as many of these as we could. Sanskrit mathematical works mentioned in the bibliography given at the end of this volume have been specially consulted by us. We are thankful to the authorities of the libraries at Madras, Bangalore, Trivandrum, Trippunithura and Benares, and those of the India Office (London) and the Asiatic Society of

Bengal (Calcutta) for supplying us transcripts of the manuscripts required or sending us manuscripts for consultation. We are indebted also to Dr. R. P. Paranjpye, Vice-Chancellor of the Lucknow University, for help in securing for our use several manuscripts or their transcripts from the state libraries in India and the India Office, London.

It would not have been possible to carry our study as far as has been done without the spade work of previous writers. Foremost among these must be mentioned the late Pandit Sudhakar Dvivedi of Benares, whose editions of the *Lîlâvatî*, *Brâhmasphuṭa-siddhânta*, *Triśatikâ*, *Mahâsiddhânta*, *Siddhânta-tattva-viveka*, etc., have been of immense help. Colebrooke's translations of the arithmetic and algebra of Brahmagupta and Bhâskara II, Kern's edition of the *Âryabhaṭîya* and Rangacarya's edition (with English translation) of the *Gaṇita-sâra-saṁgraha* of Mahâvîra have also been of much use. The recent work of G. R. Kaye, however, has been found to be extremely unreliable. His translation of the *Gaṇitapâda* of the *Âryabhaṭîya* and his edition of the Bakhshâlî Manuscript are full of mistakes and are misleading.

It has been decided to publish the book in three parts. The first part deals with the history of the numeral notation and of arithmetic. The second is devoted to algebra, a science in which the ancient Hindus made remarkable progress. The third part contains the history of geometry, trigonometry, calculus and various other topics such as magic squares, theory of series and permutations and combinations. Each part is complete in itself, so that one interested in any particular branch of mathematics need not consult all of them.

Part I which is now being published contains two chapters. Chapter I gives an account of the various devices employed by the Hindus for denoting numbers. The gradual evolution of the decimal place-value system of notation has been traced and all evidence relating to its use in India collected together for the first time. This evidence shows that the system was in use in India during the earliest centuries of the Christian era, if not earlier. We hope that the facts set forth in this chapter will finally set at rest the controversy about its place of origin. Considerations of space have prevented us from giving details regarding the introduction of the Hindu numerals into Arabia, Northern Africa and Europe. A brief account has, however, been included.

Chapter II deals with arithmetic in general. We have become so familiar with our methods of performing the fundamental arithmetical operations of addition, subtraction, multiplication, division and the extraction of roots that we seldom pause to think how and when these methods were invented. The problem, however, has deep interest for the teacher and historian of mathematics. And an account of the evolution of these methods in the land of their birth should be welcome. We have given details and illustrations of different methods of performing these operations on a *pâṭî*, ("board"), as followed in India from the fifth century onwards. It has been shown that our present methods are simple variations of those of the ancient Hindus. The rule of three, the rules of supposition and false position, and rules relating to calculations involving interest, exchange of commodities, fineness of gold, etc., are all due to the Hindus. In fact, practically the whole of elementary arithmetic can be traced back to them. Thus the importance of chapter II cannot be over-emphasised.

The scheme of transliteration of Sanskrit words

and proper names has been indicated in the beginning. Arabic words and proper names that occur in the book have been taken down as found in various secondary sources consulted.

Acknowledgment has been made in footnotes to the texts of various sources from which we have derived assistance. Of the books which have been found specially helpful, we would mention Bühler's Indian Palæography and Ojha's Prâcîna Lipi Mâlâ (in Hindi). We are indebted to Mr. Ojha for permitting us to reproduce from his book the tables of Brâhmî and Kharoṣṭhî numerals. We have pleasure in expressing our obligation to Mr. T. N. Singh for his help in the revision of proof-sheets. He has caused the removal of many obscurities and has made many valuable suggestions of which we have availed ourselves. Our thanks are also due to Mr. R. D. Misra for preparing the index to this volume.

In conclusion we express our thanks to the Law Journal Press for their unfailing courtesy and for expediting the work of printing.

LUCKNOW
September, 1935

BIBHUTIBHUSAN DATTA
AVADHESH NARAYAN SINGH

CONTENTS

CHAPTER I

NUMERAL NOTATION

CONTENTS

Chapter I

NUMERAL NOTATION

1. A GLIMPSE OF ANCIENT INDIA

The student of ancient Indian History is struck by the marvellous attainments of the Hindus, both in the Arts and the Sciences, at a very early period. The discoveries at Mohenjo-daro reveal that as early as 3,000 B.C. the inhabitants of the land of the Sindhu—the Hindus—built brick houses, planned cities, used metals such as gold, silver, copper and bronze, and lived a highly organised life. In fact, they were far in advance of any other people of that period. The earliest works available, the *Vedas* (*c.* 3,000 B.C. or probably much earlier), although consisting mainly of hymns of praise and poems of worship, show a high state of civilisation. The *Brâhmaṇa* literature (*c.* 2,000 B.C.) which follows the *Vedas*, is partly ritualistic and partly philosophical. In these works are to be found well-developed systems of metaphysical, social and religious philosophy, as well as the germs of most of the sciences and arts which have helped to make up the modern civilisation. It is here that we find the beginnings of the science of mathematics (arithmetic, geometry, algebra, etc.) and astronomy. This *Brâhmaṇa* period was followed by more than two thousand years of continuous progress and brilliant achievements. Although during this period there were several foreign invasions as well as internal wars and many great kingdoms rose and fell, yet the continuity of intellectual progress was maintained. The constitution of Hindu society was mainly responsible for this. The foreign invaders, instead of being a hindrance, contributed to progress and the strengthening of Hindu society by bringing in new blood. They settled in the land, adopted the religion and customs of the conquered and were completely absorbed into Hindu society. There were a class of people—the Brâhmaṇas—who took the vow of poverty, and devoted themselves, from one generation to another, to the cultivation of the sciences and arts, religion and philosophy. The Brâhmaṇas, thanks to their selflessness and intellectual attainments, were highly respected by the kings and the people alike. They were the law-givers and advisers of the kings. In fact, this body of selfless thinkers and learned men were the real rulers of the land.

The great Epic, the *Râmâyaṇa*, was composed by Vâlmîki, the father of Sanskrit poetry, about 1,000 B.C., Pâṇini, the grammarian, perfected Sanskrit grammar about 700 B.C. and Suśruta wrote on the sciences of medicine and surgery about 600 B.C.[1] A century later, Mahâvîra and Buddha taught their unique systems of religious and moral philosophy, and the doctrine of *Nirvâṇa*. With the spread of these religions evolved the Jaina and Buddhist literatures. Some of the earlier *Purâṇas* and *Dharma-śâstras* were written about this time. The period 400 B.C. to 400 A.D., however, seems to have been a period of great activity and progress. During this period flourished the great Jaina metaphysician Umâsvâti, Patañjali, the grammarian and philosopher, Kauṭilya, the celebrated politician, Nâgârjuna, the chemist, Caraka, the physician, and the immortal poets Aśvaghoṣa, Bhâsa and Kâlidâsa. The great

[1] There is considerable divergence of opinion regarding the dates of the pre-historic works and personalities mentioned in this section. We have given those dates that appear most plausible.

astronomical *Siddhântas*, the *Sûrya*, the *Pitâmaha*, the *Vaśiṣṭha*, the *Parâśara* and others were written during this period and the decimal place-value notation was perfected.

2. HINDUS AND MATHEMATICS

Appreciation of Mathematics. It is said that in ancient India no science did ever attain an independent existence and was cultivated for its own sake. Whatever of any science is found in Vedic India is supposed to have originated and grown as the handmaid of one or the other of the six "members of the Veda," and consequently with the primary object of helping the Vedic rituals. It is also supposed, sometimes, that any further culture of the science was somewhat discouraged by the Vedic Hindus in suspicion that it might prove a hindrance to their great quest of the knowledge of the Supreme by diverting the mind to other external channels. That is not, indeed, a correct view on the whole. It is perhaps true that in the earlier Vedic Age, sciences grew as help to religion. But it is generally found that the interest of people in a particular branch of knowledge, in all climes and times, has always been aroused and guided by specific reasons. Religion being the prime avocation of the earlier Hindus, it is not unnatural that the culture of other branches of knowledge grew as help to it and was kept subsidiary. But there is enough evidence to show that in course of time all the sciences outgrew their original purposes and were cultivated for their own sake. A new orientation had indeed set in in the latter part of the Vedic Age.

There is a story in the *Chândogya Upaniṣad*[1] whose

[1] *Chândogya Upaniṣad*, vii. 1, 2, 4.

value in support of our view cannot be over-estimated. It is said that once upon a time Nârada approached the sage Sanatkumâra and begged of him the *Brahma-vidyâ* or the supreme knowledge. Sanatkumâra asked Nârada to state what sciences and arts he had already studied so that he (Sanatkumâra) might judge what still remained to be learnt by him. Thereupon Nârada enumerated the various sciences and arts studied by him. This list included astronomy (*nakṣatra-vidyâ*) and arithmetic (*râśi-vidyâ*). Thus the culture of the science of mathematics or of any other branch of secular knowledge, was not considered to be a hindrance to spiritual knowledge. In fact, *Aparâ-vidyâ* ("secular knowledge") was then considered to be a helpful adjunct to *Parâ-vidyâ* ("spiritual knowledge").[1]

Importance to the culture of *Gaṇita* (mathematics) is also given by the Jainas. Their religious literature is generally classified into four branches, called *anuyoga* ("exposition of principles"). One of them is *gaṇitânu-yoga* ("the exposition of the principles of mathematics"). The knowledge of *Saṁkhyâna* (literally, "the science of numbers," meaning arithmetic and astronomy) is stated to be one of the principal accomplishments of the Jaina priest.[2] In Buddhist literature too, arithmetic (*gaṇanâ*, *saṁkhyâna*) is regarded as the first and the noblest of the arts.[3] All these will give a fair idea of the importance and value set upon the culture of *gaṇita* in ancient India.

The following appreciation of mathematics, although belonging to a much later date, will be found to be interesting, especially, as it comes from the pen

[1] *Muṇḍakopaniṣad*, i. 1, 3-5.

[2] *Bhagavatî-sûtra*, Sûtra 90; *Uttarâdhyayana-sûtra*, xxv. 7, 8, 38.

[3] *Vinaya Piṭaka*, ed. Oldenberg, Vol. IV, p. 7; *Majjhima Nikâya*, Vol. I, p. 85; *Cullaniddesa*, p. 199.

of Mahâvîra (850 A.D.), one of the best mathematicians of his time:

"In all transactions which relate to worldly, Vedic or other similar religious affairs calculation is of use. In the science of love, in the science of wealth, in music and in drama, in the art of cooking, in medicine, in architecture, in prosody, in poetics and poetry, in logic and grammar and such other things, and in relation to all that constitutes the peculiar value of the arts, the science of calculation (*gaṇita*) is held in high esteem. In relation to the movements of the sun and other heavenly bodies, in connection with eclipses and conjunctions of planets, and in connection with the *tripraśna* (direction, position and time) and the course of the moon—indeed in all these it is utilised. The number, the diameter and the perimeter of islands, oceans and mountains; the extensive dimensions of the rows of habitations and halls belonging to the inhabitants of the world, of the interspace between the worlds, of the world of light, of the world of the gods and of the dwellers in hell, and other miscellaneous measurements of all sorts—all these are made out by the help of *gaṇita*. The configuration of living beings therein, the length of their lives, their eight attributes, and other similar things; their progress and other such things, their staying together, etc.—all these are dependent upon *gaṇita* (for their due comprehension). What is the good of saying much? Whatever there is in all the three worlds, which are possessed of moving and non-moving beings, cannot exist as apart from *gaṇita* (measurement and calculation).

" With the help of the accomplished holy sages, who are worthy to be worshipped by the lords of the world, and of their disciples and disciples' disciples, who constitute the well-known series of preceptors,

I glean from the great ocean of the knowledge of numbers a little of its essence, in the manner in which gems are picked from the sea, gold is from the stony rock and pearl from the oyster shell; and give out according to the power of my intelligence, the *Sâra-saṁgraha*, a small work on *gaṇita*, which is (however) not small in value."[1]

Mathematics in Hindu Education. The elementary stage in Hindu education lasted from the age of five till the age of twelve. This period slightly differed in the case of sons of kings and noblemen. The main subjects of study were *lipi* or *lekhâ* (alphabets, reading and writing), *rûpa* (drawing and geometry) and *gaṇanâ* (arithmetic). It is said in the *Arthaśâstra* of Kauṭilya (400 B.C.) that having undergone the ceremony of tonsure, the student shall learn the alphabets (*lipi*) and arithmetic (*saṁkhyâna*).[2] We find in the Hâthîgumphâ Inscription[3] that king Khâravela (163 B.C.) of Kaliṅga spent nine years (from the age of sixteen to the age of 25) in learning *lekhâ*, *rûpa* and *gaṇanâ*. Prince Gautama began his education when he was eight years of age "firstly (with) writing and then arithmetic as the most important of the 72 sciences and arts."[4] Mention of *lekhâ*, *rûpa* and *gaṇanâ* is also found in the Jaina canonical works.[5]

[1] *GSS*, i. 9-19.

[2] *Arthaśâstra*, ed. by R. Shamasastri, i. 5, 2; Eng. trans., p. 10.

[3] *Hathigumpha and three other inscriptions*, ed. by Bhagavanlal Indraji, p. 22.

[4] *Antagaḍa-dasâo and Anuttaravavâiya-dasâo*, Eng. trans. by L. D. Barnett, 1907, p. 30; cf. *Kalpasûtra* of Bhadrabâhu, Sûtra 211.

[5] E.g., *Samavâyâṅga-sûtra*, Sûtra 72.

3. SCOPE AND DEVELOPMENT OF HINDU MATHEMATICS

Gaṇita literally means "the science of calculation" and is the Hindu name for mathematics. The term is a very ancient one and occurs copiously in Vedic literature. The *Vedâṅga Jyotiṣa* (*c.* 1,200 B.C.) gives it the highest place of honour among the sciences which form the *Vedâṅga*: "As the crests on the heads of peacocks, as the gems on the hoods of snakes, so is *gaṇita* at the top of the sciences known as the *Vedâṅga*."[1] In ancient Buddhist literature we find mention of three classes of *gaṇita*: (1) *mudrâ* ("finger arithmetic"), (2) *gaṇanâ* ("mental arithmetic") and (3) *saṁkhyâna* ("higher arithmetic in general"). One of the earliest enumerations of these three classes occurs in the *Dîgha Nikâya*,[2] and it is also found in the *Vinaya Piṭaka*,[3] *Divyâvadâna*[4] and *Milindapañho*.[5] The word *saṁkhyâna* has been used for *gaṇita* in several old works.[6] At this remote period *gaṇita* included astronomy, but geometry (*kṣetra-gaṇita*) belonged to a different group of sciences known as *Kalpasûtra*.

It is believed that some time before the beginning of the Christian era, there was a renaissance of Hindu *Gaṇita*.[7] The effect of this revival on the scope of

[1] "Yathâ śikhâ mayurâṇâṁ nâgânâṁ maṇayo yathâ
Tadvadvedâṅgaśâstrâṇâṁ gaṇitaṁ mûrdhani sthitaṃ."
Vedâṅga Jyotiṣa, 4.

[2] I, p. 51.

[3] IV, p. 7.

[4] *Divyâvadâna*, ed. by E. B. Cowell and R. A. Neil, Cambridge, 1886, pp. 3, 26 and 88.

[5] *Milindapañho*, Eng. trans. by Rhys Davids, Oxford, 1890, p. 91.

[6] E.g., *Kalpasûtra* of Bhadrabâhu, ed. by H. Jacobi, Leipzig, 1897; *Bhagavatî-sûtra*, Bombay, 1918, p. 112; *Arthaśâstra*, i. 5. 2.

[7] Bibhutibhusan Datta, "The scope and development of Hindu Gaṇita," *Indian Historical Quarterly*, V, 1929, pp. 479-512.

gaṇita was great. Astronomy (*jyotiṣa*) became a separate subject and geometry (*kṣetra-gaṇita*) came to be included within its scope. The subjects treated in the Hindu *Gaṇita* of the early renaissance period consisted of the following:[1] *Parikarma* ("fundamental operations"), *Vyavahâra* ("determinations"), *Rajju* ("rope," meaning geometry), *Râśi* ("rule of three"), *Kalâsavarṇa* ("operations with fractions"), *Yâvat tâvat* ("as many as," meaning simple equations), *Varga* ("Square," meaning quadratic equations), *Ghana* ("Cube", meaning cubic equations), *Varga-varga* (biquadratic equations) and *Vikalpa* ("permutations and combinations").

Thus *gaṇita* came to mean mathematics in general, while 'finger arithmetic' as well as 'mental arithmetic' were excluded from the scope of its meaning For the calculations involved in *gaṇita*, the use of some writing material was essential. The calculations were performed on a board (*pâṭî*) with a piece of chalk or on sand (*dhûlî*) spread on the ground or on the *pâṭî*. Thus the terms *pâṭî-gaṇita* ("science of calculation on the board") or *dhûlî-karma* ("dust-work"), came to be used for higher mathematics. Later on the section of *gaṇita* dealing with algebra was given the name *Bîja-gaṇita*. The first to effect this separation was Brahmagupta (628), but he did not use the term *Bîja-gaṇita*. The chapter dealing with algebra in his *Brâhma-sphuṭa-siddhânta* is called *Kuṭṭaka*. Śrîdharâcârya (750) regarded *Pâṭî-gaṇita* and *Bîja-gaṇita* as separate and wrote separate treatises on each. This distinction between *Pâṭîgaṇita* and *Bîjagaṇita* has been preserved by later writers.

Having given a brief survey of the position and scope of mathematics in Ancient India, we turn to the

[1] "Parikammaṁ vavahâro rajju râsî kalâsavanne ya ।
Jâvantâvati vaggo ghano tataha vaggavaggo vikappo ta ॥"
Sthânâṅgasûtra, Sûtra 747.

purpose in hand—that of giving a connected account of the development and growth of the different branches of mathematics. The numeration system of the Hindus will engage our attention first.

4. NUMERAL TERMINOLOGY

Scale of Notation. We can definitely say that from the very earliest known times, ten has formed the basis of numeration in India.[1] In fact, there is absolutely no trace of the extensive use of any other base of numeration in the whole of Sanskrit literature. It is also characteristic of India that there should be found at a very early date long series of number names for very high numerals. While the Greeks had no terminology for denominations above the *myriad* (10^4), and the Romans above the *mille* (10^3), the ancient Hindus dealt freely with no less than eighteen denominations. In modern times also, the numeral language of no other nation is as scientific and perfect as that of the Hindus.

In the *Yajurveda Saṁhitâ* (*Vâjasaneyî*)[2] the following list of numeral denominations is given: *Eka* (1), *daśa* (10), *sata* (100), *sahasra* (1000), *ayuta* (10,000), *niyuta* (100,000), *prayuta* (1,000,000), *arbuda* (10,000,000), *nyarbuda* (100,000,000), *samudra* (1,000,000,000), *madhya* (10,000,000,000), *anta* (100,000,000,000), *parârdha* (1,000,000,000,000). The same list occurs at two places in the *Taittirîya Saṁhitâ*.[3] The *Maitrâyaṇî*[4]

[1] Various instances are to be found in the *Ṛgveda;* noted by Macdonell and Keith, *Vedic Index*, Vol. I, p. 343.

[2] *Yajurveda Saṁhitâ*, xvii. 2.

[3] iv. 40. 11. 4; and vii. 2. 20. 1.

[4] ii. 8. 14; the list has *ayuta*, *prayuta*, then again *ayuta*, then *nyarbuda*, *samudra*, *madhya*, *anta*, *parârdha*.

and *Kâṭhaka*[1] *Saṁhitâs* contain the same list with slight alterations. The *Pañcaviṁśa Brâhmaṇa* has the *Yajurveda* list upto *nyarbuda* inclusive, and then follow *nikharva*, *vâdava*, *akṣiti*, etc. The *Sâṅkhyâyana Srauta Sûtra* continues the series after *nyarbuda* with *nikharva*, *samudra*, *salila*, *antya*, *ananta* (= 10 billions). Each of these denominations is 10 times the preceding, so that they were aptly called *daśaguṇottara saṁjñâ*[2] ("decuple terms").

Coming to later times, *i.e.*, about the 5th century B.C., we find successful attempts made to continue the series of number names based on the centesimal scale.[3] We quote below from the *Lalitavistara*,[4] a well-known Buddhist work of the first century B.C., the dialogue between Arjuna, the mathematician, and Prince Gautama (Bodhisattva):

"The mathematician Arjuna asked the Bodhisattva, 'O young man, do you know the counting which goes beyond the *koṭi* on the centesimal scale?

Bodhisattva: I know.

Arjuna: How does the counting proceed beyond the *koṭi* on the centesimal scale?

Bodhisattva: Hundred *koṭis* are called *ayuta*, hundred *ayutas niyuta*, hundred *niyutas kaṅkara*, hundred *kaṅkaras vivara*, hundred *vivaras kṣobhya*, hundred *kṣobhyas vivâha*, hundred *vivâhas utsaṅga*, hundred *utsaṅgas bahula*, hundred *bahulas nâgabala*, hundred *nâgabalas tiṭi-*

[1] xvii. 10; the list is the same with the exception that *niyuta* and *prayuta* change places. In xxxix. 6, after *nyarbuda* a new term *vâdava* intervenes.

[2] Cf. Bhâskara II, L, p. 2.

[3] *Śatottara gaṇanâ* or *Śatottara saṁjñâ* (names on the centesimal scale).

[4] *Lalitavistara*, ed. by Rajendra Lal Mitra, Calcutta, 1877,

lambha, hundred *tiṭilambhas vyavasthâna-prajñapti*, hundred *vyavasthâna-prajñaptis hetuhila*, hundred *hetuhilas karahu*, hundred *karahus hetvindriya*, hundred *hetvindriyas samâpta-lambha*, hundred *samâpta-lambhas gaṇanâgati*, hundred *gaṇanâgatis niravadya*, hundred *niravadyas mudrâ-bala*, hundred *mudrâ-balas sarva-bala*, hundred *sarva-balas visaṁjñâ-gati*, hundred *visaṁjñâ-gatis sarvajñâ*, hundred *sarvajñâs vibhutaṅgamâ*, hundred *vibhutaṅgamâs tallakṣaṇa*.[1]"

Another interesting series of number names increasing by multiples of 10 millions is found in Kâccâyana's Pali Grammar.[2] "For example: *dasa* (10) multiplied by *dasa* (10) becomes *sata* (100), *sata* (100) multiplied by ten becomes *sahassa* (1,000), *sahassa* multiplied by ten becomes *dasa sahassa* (10,000), *dasa sahassa* multiplied by ten becomes *sata sahassa*[3] (100,000), *sata sahassa* multiplied by ten becomes *dasa sata sahassa* (1,000,000), *dasa sata sahassa* multiplied by ten becomes *koṭi* (10,000,000). Hundred-hundred-thousand *koṭis* give *pakoṭi*.[4] In this manner the further terms are formed. What are their names? .... hundred hundred-thousands is *koṭi*, hundred-hundred-

[1] Thus *tallakṣaṇa* $=10^{53}$.

This and the following show that the Hindus anticipated Archimedes by several centuries in the matter of evolving a series of number names which "are sufficient to exceed not only the number of a sand-heap as large as the whole earth, but one as large as the universe."

Cf. '*De harenae numero*' in the 1544 edition of the *Opera* of Archimedes; quoted by Smith and Karpinski, *Hindu Arabic Numerals*, Boston, 1911, p. 16.

[2] "Grammaire Pâlie de Kâccâyana," *Journ. Asiatique*, Sixieme Serie, XVII, 1871, p. 411. The explanations to sûtras 51 and 52 are quoted here.

[3] Also called *lakkha* (*lakṣa*).

[4] Also called *koṭi-koṭi*, i.e., $(10,000,000)^2=10^{14}$. The following numbers are in the denomination *koṭi-koṭi*. Compare the *Anuyogadvâra-sûtra*, Sûtra 142.

thousand *koṭis* is *pakoṭi*, hundred-hundred-thousand *pakoṭis* is *koṭippakoṭi*, hundred-hundred-thousand *koṭippakoṭis* is *nahuta*, hundred-hundred-thousand *nahutas* is *ninnahuta*, hundred-hundred-thousand *ninnahutas* is *akkhobhini*; similarly we have *bindu*, *abbuda*, *nirabbuda*, *ahaha*, *ababa*, *atata*, *sogandhika*, *uppala*, *kumuda*, *puṇḍarîka paduma*, *kathâna*, *mahâkathâna*, *asaṅkhyeya*."[1]

In the *Anuyogadvâra-sûtra*[2] (*c.* 100 B.C.), a Jaina canonical work written before the commencement of the Christian era, the total number of human beings in the world is given thus: "a number which when expressed in terms of the denominations, *koṭi-koṭi*, etc., occupies twenty-nine places (*sthâna*), or it is beyond the 24th place and within the 32nd place, or it is a number obtained by multiplying sixth square (of two) by (its) fifth square, (*i.e.*, 2^{96}), or it is a number which can be divided (by two) ninety-six times." Another big number that occurs in the Jaina works is the number representing the period of time known as *Śîrṣaprahelikâ*. According to the commentator Hema Candra (b. 1089)[3], this number is so large as to occupy 194 notational places (*aṅka-sthânehi*). It is also stated to be $(8{,}400{,}000)^{28}$.

Notational Places. Later on, when the idea of place-value was developed, the denominations (number names) were used to denote the places which unity would occupy in order to represent them (denominations) in writing a number on the decimal scale. For instance, according to Âryabhaṭa I (499) the denominations are the names of 'places'. He says: "*Eka* (unit) *daśa* (ten), *śata* (hundred), *sahasra* (thousand), *ayuta* (ten thousand), *niyuta* (hundred thousand), *prayuta* (million),

[1] Thus *asaṅkhyeya* is $(10)^{140} = (10{,}000{,}000)^{20}$.

[2] Sûtra 142.

[3] The figures within brackets after the names of authors or works denote dates after Christ.

koṭi (ten million), *arbuda* (hundred million), and *vṛnda* (thousand million) are respectively from *place to place* each ten times the preceding."[1] The first use of the word 'place' for the denomination is met with in the Jaina work quoted above.

In most of the mathematical works, the denominations are called "names of places," and eighteen of these are generally enumerated. Śrîdhara (750) gives the following names:[2] *eka*, *daśa*, *śata*, *sahasra*, *ayuta*, *lakṣa*, *prayuta*, *koṭi*, *arbuda*, *abja*, *kharva*, *nikharva*, *mahâsaroja*, *śaṅku*, *saritâ-pati*, *antya*, *madhya*, *parârdha*, and adds that the decuple names proceed even beyond this. Mahâvîra (850) gives twenty-four notational places:[3] *eka*, *daśa*, *śata*, *sahasra*, *daśa-sahasra*, *lakṣa*, *daśa-lakṣa*, *koṭi*, *daśa-koṭi*, *śata-koṭi*, *arbuda*, *nyarbuda*, *kharva*, *mahâkharva*, *padma*, *mahâ-padma*, *kṣoṇi*, *mahâ-kṣoṇi*, *śaṅkha*, *mahâ-śaṅkha*, *kṣiti*, *mahâ-kṣiti*, *kṣobha*, *mahâ-kṣobha*.

Bhâskara II's (1150) list agrees with that of Śrîdhara except for *mahâsaroja* and *saritâpati* which are replaced by their synonyms *mahâpadma* and *jaladhi* respectively. He remarks that the names of places have been assigned for practical use by ancient writers.[4]

Nârâyaṇa (1356) gives a similar list in which *abja*, *mahâsaroja* and *saritâpati* are replaced by their synonyms *saroja*, *mahâbja* and *pârâvâra* respectively.

Numerals in Spoken Language. The Sanskrit names for the numbers from one to nine are: *eka*, *dvi*, *tri*, *catur*, *pañca* *ṣaṭ*, *sapta*, *aṣṭa*, *nava*. These with the

[1] *Â*, ii. 2.

[2] *Triś*, R. 2-3; the term used is *daśaguṇâḥ saṁjñâḥ*, i.e., "decuple names."

[3] *GSS*, i. 63-68; "The first place is what is known as *eka*; the second is *daśa*" etc.

[4] *L*, p. 2.

numerical denominations already mentioned suffice to express any required number. In an additive system it is immaterial how the elements of different denominations, of which a number is composed, are spoken. Thus one-ten or ten-one would mean the same. But it has become the usual custom from times immemorial to adhere to a definite mode of arrangement, instead of speaking in a haphazard manner.

In the Sanskrit language the arrangement is that when a number expression is composed of the first two denominations only, the smaller element is spoken first, but when it is composed also of higher denominations, the bigger elements precede the smaller ones, the order of the first two denominations remaining as before. Thus, if a number expression contains the first four denominations, the normal mode of expression would be to say the thousands first, then hundreds, then units and then tens. It will be observed that there is a sudden change of order in the process of formation of the number expression when we go beyond hundred. The change of order, however, is common to most of the important languages of the world.[1] Nothing definite appears to be known as to the cause of this sudden change.

The numbers 19, 29, 39, 49, etc., offer us instances of the use of the *subtractive* principle in the spoken language. In Vedic times we find the use[2] of the terms *ekânna-vimśati* (one-less-twenty) and *ekânna-catvârimśat* (one-less-forty) for nineteen and thirty-nine respectively. In later times (Sûtra period) the *ekânna* was changed to *ekona*, and occasionally even the prefix *eka*

[1] Only in very few languages is the order continuously descending. In English the smaller elements are spoken first in the case of numbers upto twenty only.

[2] *Taittiriya Samhitâ*, vii. 2. 11.

was deleted and we have *ûna-vimśati*, *ûna-trimśat*, etc.—forms which are used upto the present day. The alternative expressions *nava-daśa* (nine-ten), *nava-vimśati* (nine-twenty), etc., were also sometimes used.[1]

Practically the whole of Sanskrit literature is in verse, so that for the sake of metrical convenience, various devices were resorted to in the formation of number expressions, the most common being the use of the additive[2] method. The following are a few examples of common occurrence taken from mathematical works:

Subtractive: (1) the number 139 is expressed as $40+100-1$;[3]

(2) 297 is expressed as $300-3$.[4]

Multiplicative: (1) the number eighteen is expressed as 2×9;[5]

(2) twenty-seven is expressed as 3×9 and 12 as 2×6;[6]

(3) 28,483 is expressed as $83+400+(4000\times7)$.

[1] 19 = *nava-daśa* (*Vâjasaneyî Samhitâ*, xiv. 23; *Taittirîya Samhitâ*, xiv. 23. 30).
29 = *nava-vimśati* (*Vâjasaneyî Samhitâ*, xiv. 31).
99 = *nava-navati* (*Ṛgveda*, i. 84. 13).

[2] 3339 = *trîṇi śatâni trisahasrâṇi trimśa ca nava ca*, i.e., "three hundreds and three thousands and thirty and nine." (*Ṛgveda*, iii. 9. 9; also x. 52. 6.)

[3] *GSS*, i. 4: *cattvarimśaścaikona śatâdhika* ("forty increased by one-less-hundred").

[4] *L*, p. 4, *Ex*. 1: *Trihînasya śata-trayasya* ("three less three hundred").

[5] *Â*, ii. 3: *dvi-navaka*.

[6] *Triś*, *Ex*. 43: *tri-navaka* ("three nines"), *dvi-ṣaṭ* ("two sixes").

[7] *GSS*, i. 28: *tryaśîtimiśrâṇi catuśśatâni catussahasraghna nagânvitâni* ("eighty-three combined with four hundred and four thousand multiplied by seven").

The expression of the number 12345654321 in the form "beginning with 1 upto 6 and then diminishing in order" is rather interesting.[1]

What are known as alphabetic and word numerals were generally employed for the expression of large numbers. A detailed account of these numerals will be given later on.

5. THE DEVELOPMENT OF NUMERICAL SYMBOLISM

Writing in Ancient India. It is generally held that numerical symbols were invented after writing had been in use for some time, and that in the early stages the numbers were written out in full in words. This seems to be true for the bigger units, but the signs for the smaller units are as old as writing itself.

Until quite recently historians were divided as to the date when writing was in use in India. There were some who stated that writing was known even in the Vedic age, but the majority following Weber, Taylor, Bühler and others were of opinion that writing was introduced into India from the West about the eighth century B.C. These writers built up theories deriving the ancient Indian script, as found in the inscriptions of Aśoka, from the more ancient writing discovered in Egypt and Mesopotamia. The Semitic origin was first suggested by Sir W. Jones, in the year 1806, and later on supported by Kopp (1821), Lespius (1834), and many others. The supporters of this theory, however, do not completely agree amongst themselves. For, whilst W. Deecke and I. Taylor derive the Indian script from a South-Semitic script, Weber and Bühler derive

[1] *GSS*, i.27: *ekâdiṣaḍantâni krameṇa hînâni.*

it from the Phoenician or a North-Semitic script.[1] Bühler rejects the derivation from a South-Semitic script, stating that the theory requires too many assumptions, and makes too many changes in the letter forms to be quite convincing. He, however, supports Weber's derivation from a North-Semitic script and has given details of the theory.[2] Ojha[3] has examined Bühler's theory in detail and rejects it stating that it is fanciful and that the facts are against it. He states that only one out of the twenty-two letters of the Phoenician (North-Semitic) script resembles a phonetically similar Brâhmî letter. He supports his argument in a most convincing manner by a table of the two alphabets, with phonetically similar letters arranged in a line. He further shows that following Bühler's method of derivation almost any script could be proved to be the parent of another.[4]

Other scholars, who held that writing was known in India as early as the Vedic age, based their conclusion upon literary evidence. The *Vaśiṣṭha Dharmasûtra*, which originally belonged to a school of the *Ṛgveda* offers clear evidence of the use of writing in the Vedic period. Vaśiṣṭha (xvi. 10, 14-15) mentions written documents as legal evidence, and the first of these sûtras

[1] For minor differences in the theories set up by different writers and also for several other theories, see Bühler, *Palaeography*, p. 9; the notes give the references.

[2] Bühler, *l.c.*, pp. 9f.

[3] *PLM*, pp. 18-31.

[4] Recently several other eminent historians have expressed their disagreement with Bühler's derivation. See Bhandarkar, "Origin of the Indian Alphabet," *Sir Asutosh Mukerji Jubilee Volumes*, Vol. III, 1922, p. 493; H. C. Ray, "The Indian Alphabet," *IA*, III, 1924, p. 233; also *Mohenjo-daro and the Indus Valley Civilisation*, 1931, p. 424, where the following remark occurs: "I am convinced that all attempts to derive the Brâhmî alphabet from Semitic alphabets were complete failures."

is a quotation from an older work or from traditional lore. Another quotation from the *Ṛgveda* itself (x. 62. 7), which refers to the writing of the number eight is: *Sahasraṁ me dadato -aṣṭakarṇyaḥ*, meaning "gave me a thousand cows on whose ears the number eight was written." The above interpretation, although doubted by some scholars, seems to be correct, as it is supported by Pâṇini.[1] Moreover, the practice of making marks on the ears of cows to denote their relation to their owners, seems to have been prevalent in ancient India.[2]

At another place in the *Ṛgveda* (x. 34), we find mention of a gambler lamenting his lot and saying that "having staked on one,[3] he lost his faithful wife...." Again, in the *Atharvaveda* (vii. 50, (52), 5) we find the mention of the word "written amount".[4] Pâṇini's grammar (*c.* 700 B.C.) contains the terms *yavanâni* ("Semitic writing") and the compounds *lipikâra* and *libikâra* (iii. 2. 21) (writer), which show that writing was known in his time. In addition to these passages, the Vedic works contain some technical terms, such as *akṣara* (a letter of the alphabet), *kâṇḍa* (chapter), *paṭala*, *grantha* (book), etc., which have been quoted as evidence of writing. These specific references to written documents when considered with the advanced state of Vedic civilisation, especially the high development of trade and complicated monetary transactions, the use of prose in the *Brâhmaṇas*, the collection, the methodical arrangement, the numeration, the analysis of the Vedic texts

[1] *Karṇo varṇa lakṣaṇât* (vi. 2. 112) and also (vi. 3. 115) support the interpretation.

[2] *Atharvaveda* (vi. 141) mentions the method of making *mithuna* marks on the ears. In (xii. 4. 6) the practice is denounced. The *Maitrâyaṇî Saṁhitâ* has a chapter dealing with this topic. The method of making such marks is dealt with in iv. 2. 9.

[3] Here 'one' refers to the number stamped on the dice.

[4] *Ajaiṣaṃ tvâ saṁlikhitamajaiṣamuta saṁrudhaṃ.*

and the phonetic and lexicographic researches found in the *Vedâṅgas*, form sufficient grounds for assigning a very early date to the use of writing in India.[1] Although these arguments possess considerable weight, they were not generally recognised, as will always happen if an *argumentum ex impossibili* is used. R. Shamasastry (1906) has published a derivation based upon ancient Indian hieroglyphic pictures which he believes to be preserved in the *tântric* figures. His learned article has not attracted the attention it deserves.

Recent discoveries have however, sounded the death knell of all theories deriving the Indian script from foreign sources. Pottery belonging to the Megalithic (*c.* 1,500 B.C.) and Neolithic (6,000 B.C.—3,000 B.C.) ages, preserved in the Madras Museum, has been found to be inscribed with writing. And according to Bhandarkar[2] five of these marks are identical with the Brâhmî characters of the time of Aśoka. The excavations at Mohenjo-daro and Harappa have also brought to light written documents, seals and inscriptions, dating from before 3,000 B.C. Thus it would be now absurd to trace the Brâhmî to any Semitic alphabet of the eighth or ninth century B.C.

Earliest Numerals. The numerical figures contained in the seals and inscriptions of Mohenjo-daro, have not been completely deciphered as yet. The vertical stroke and combinations of vertical strokes arranged side by side, or one group below another, have been found. The numbers one to thirteen seem to have been written by means of vertical strokes, probably, as in the figures given below:[3]

[1] Cf. Bühler, *l.c.*, p. 3. [2] *l.c.*

[3] Marshall, *l.c.*, pp. 450-52. See also "Mohenjo-daro—Indus Epigraphy" by G. R. Hunter (*JRAS*, April, 1932, pp. 470, 478ff.) who is more pronounced about the numerical values of some of the signs.

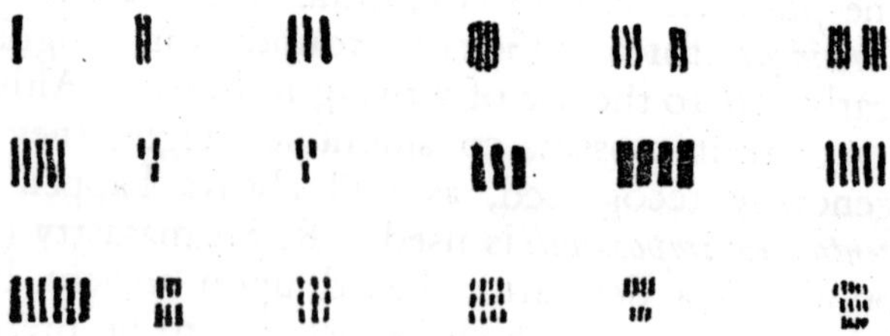

It is not yet quite certain whether there were special signs for greater numbers such as 20, 30, the hundreds and higher numbers. There are numerous other signs which are believed to represent such numbers, but there seems to be no means of finding out the true values of these signs at present.

Between the finds of Mohenjo-daro and the inscriptions of Aśoka, which contain numerals, there is a gap of 2,700 years or more. No written documents containing numerals and belonging to this intervening period have been so far discovered. The literary evidence, however, points to the use of numerical symbols at a very early date. The reference to the writing of the number eight in the *Ṛgveda* and the use of numerical denominations as big as 10^{12} in the *Yajurveda Saṁhitâ* and in several other Vedic works, quoted before, offer sufficient grounds for concluding that, even at that remote period, the Hindus must have possessed a well developed system of numerical symbols. The conclusion is supported by the fact that the Greek and the Roman numerical terminologies did not go beyond 10^4, even after writing and a satisfactory numerical symbolism had been in use for several centuries.

The writings on the inscriptions of Aśoka show that in his time the use of numerical symbols in India was quite common.[1] The variations in the forms of the numerical signs suggest that the symbols had been in use for a long time.

Most of the inscriptions of Aśoka and the following period are written in a script which has been called *Brâhmî*, whilst some are in a different script known as *Kharoṣṭhî*. The forms of the numerical symbols in the two scripts are different. We consider them separately.

6. KHAROṢṬHÎ NUMERALS

Early Occurrence. The *Kharoṣṭhî lipi* is a script written from right to left. The majority of the Kharoṣṭhî inscriptions have been found in the ancient province of Gândhâra, the modern eastern Afghanistan and the northern Punjab. It was a popular script meant for clerks and men of business. The period during which it seems to have been used in India extends from the fourth century B.C. to the third century A.D. In the Kharoṣṭhî inscriptions of Aśoka only four numerals have been found. These are the primitive vertical marks for one, two, four, and five, thus:

1 2 4 5

More developed forms of these numerals are found in the inscriptions of the Sakas, of the Pârthians and

[1] Megasthenes speaks of mile-stones indicating the distances and the halting places on the roads. The distances must have been written in numerical figures (Bühler, *l.c.*, p. 6; also *Indika of Megasthenes*, pp. 125-26). The complicated system of keeping accounts mentioned in the *Arthaśâstra* of Kauṭilya confirms the conclusion.

of the Kuṣânas, of the 1st century B.C. and the 1st and 2nd centuries A.D., as well as in other probably later documents. The following are some of the numerals of this period:

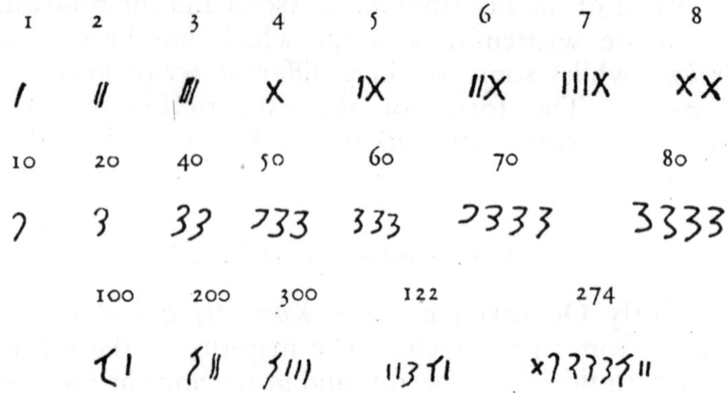

Forms and their Origin. It cannot be satisfactorily explained why the number four, which was previously represented by four vertical lines came to be represented by a cross later on. The representation of the numbers five to eight follows the additive principle, with four as the base. This method of writing the numbers 4 to 8 is not met with in the early records of the Semites. We do not know how the number nine was written. It is very probable that it was written as IXX, *i.e.*, 4+4+1 (reading from right to left, the order being the same as that of the script). The number 10 has an entirely new sign. The question why it was not written as IIXX, or why the base X (4) was abandoned cannot be satisfactorily answered.

It is accepted by all that the Kharoṣṭhî is a foreign script brought into India from the west. The exact period at which it was imported is unknown. It might have been introduced at the time of the conquest of the Punjab by Darius (*c.* 500 B.C.) or earlier.[1] The numerals given above undoubtedly belong to this script as they proceed from right to left.

The old symbols of the inscriptions of Aśoka, however, seem to have undergone modification in India, especially the numbers from 4 to 19. The symbols for four and ten seem to have been coined in India, in order to introduce simplification and also to bring the Kharoṣṭhî numeral system in line with the Brâhmî notation already in extensive use. The symbol X seems to have been derived by turning the Brâhmî symbol + which represents 4 in the inscriptions of Aśoka. The inclined cross to represent 4 is found in the Nabatean numerals in use in the earlier centuries of the Christian Era.[2] The Nabatean numerals resemble the Kharoṣṭhî also in the use of the scale of twenty and in the method of formation of the hundreds. It is possible that the Semites might have borrowed the Kharoṣṭhî symbol for 4, although it is not unlikely, as Bühler thinks, that the symbol might have been invented independently by both nations.

[1] The theory of the foreign origin of the script has to be revised in the light of the discoveries at Mohenjo-daro and Harappa, especially in view of the fact that the Mohenjo-daro alphabet ran from right to left.

[2] J. Euting, *Nabatäische Inschriften aus Arabien*, Berlin, 1885, pp. 96-97.

The numeral  (10) closely resembles the letter *a* of the Brâhmî alphabet. The symbol for twenty  appears to be a cursive combination of two tens. It resembles one of the early Phoenician forms found in the papyrus Blacas[1] (5th century B.C.). The mode of expressing the numbers 30, 40, etc., by the help of the symbols for 10 and 20, is the same as amongst the early Phoenicians and Aramaeans.

The symbol for 100 resembles the letter *ta* or *tra* of the Brâhmî script, to the right of which stands a vertical stroke.

The symbols for 200, 300, etc., are formed by writing the symbols for 2, 3, etc., respectively to the right of the symbol for 100. This evidently is the use of the multiplicative principle, as is found amongst the early Phoenicians.[2]

The formation of other numbers may be illustrated by the number 274 which is written with the help of the symbols for 2, 100, 20, 10 and 4 arranged as

in the right to left order. The 2 on the right of 100 multiplies 100, whilst the numbers written to the left are added, thus giving 274.

The ancient Kharoṣṭhî numerals are given in Table I.

[1] Bühler, *Palaeography*, p. 77; Ojha, *l.c.*, p. 128; see Table II(*b*).
[2] See Table II(*c*).

7. BRÂHMÎ NUMERALS

Early Occurrence and Forms. The Brâhmî inscriptions are found distributed all over India. The Brâhmî script was, thus, the national script of the ancient Hindus. It is undoubtedly an invention of the Brâhmaṇas. The early grammatical and phonetic researches seem to have resulted in the perfection of this script about 1,000 B.C. or earlier. The Brâhmî numerals are likewise a purely Indian invention. Attempts have been made by several writers of note to evolve a theory of a foreign origin of the numerals, but we are convinced that all those attempts were utter failures.[1] These theories will be dealt with at their proper places. Due to the lack of early documents, we are not in a position to say what exactly were the original forms of the Brâhmî symbols. Our knowledge of these symbols goes back to the time of King Aśoka (*c.* 300 B.C.) whose vast dominions included the whole of India and extended in the north upto Central Asia. The forms of these symbols are:

4 6 50 200

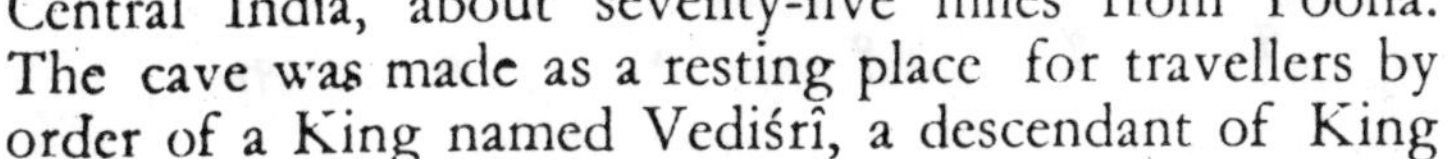

The next important inscription containing numerals is found in a cave on the top of the Nânâghât hill in Central India, about seventy-five miles from Poona. The cave was made as a resting place for travellers by order of a King named Vediśrî, a descendant of King Sâtavâhana. The inscription contains a list of gifts made on the occasion of the performance of several *yajñas* or religious sacrifices. It was first deciphered

[1] *Cf.* Langdon's opinion in *Mohenjo-daro and the Indus Valley Civilisation*, ch. xxiii.

by Paṇdit Bhagavanlal Indraji who has given the interpretation of the numerical symbols.[1] These occur at about thirty places, and their forms are as below:

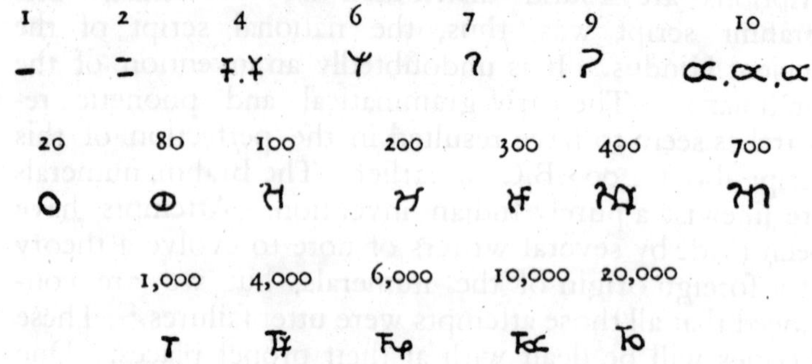

A number of inscriptions containing numerals and dating from the first or the second century A.D. are found in a cave in the district of Nasik in the Bombay presidency. These contain a fuller list of numerals. The forms[2] are as follows:

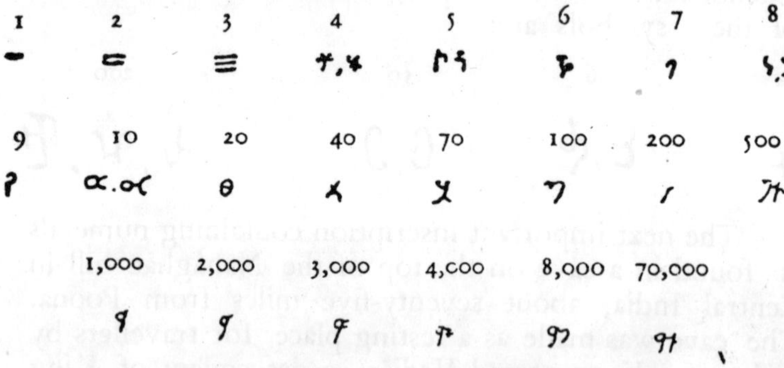

[1] "On Ancient Nâgarî Numeration; from an inscription at Nânâghât," *Journ. of the Bombay Branch of the Royal Asiatic Society*, 1876, Vol. XII, p. 404.

[2] E. Senart, "The inscriptions in the caves at Nasik," *EI*, Vol. VIII, pp. 59-96; "The inscriptions in the cave at Karle," *EI*, Vol. VII, pp. 47-74.

Even after the invention of the zero and the place-value system, the same numerical symbols from 1 to 9, continued to be employed with the zero to denote numbers. Thus the gradual development of these forms can be easily traced. This gradual change from the old system without place-value to the new system with the zero and the place-value is to be met with in India alone. All other nations of the world have given up their indigenous numerical symbols which they had used without place-value and have adopted the zero and a new set of symbols, which were never in use in those countries previously. This fact alone is a strong proof of the Hindu origin of the zero and the place-value system.

The numbers 1, 2 and 3 of the Brâhmî notation were denoted by one, two and three horizontal[1] lines placed one below the other. These forms clearly distinguish the Brâhmî notation from the Kharoṣṭhî and the Semitic systems.

It cannot be said why the strokes were horizontal in Brâhmî and vertical in Kharoṣṭhî and Semitic writings, just as it cannot be said why the writing proceeded from left to right in Brâhmî and from right to left in Kharoṣṭhî and Semitic writings. It appears to us that the Brâhmî and the Kharoṣṭhî (Semitic) numerals have always existed side by side and it cannot be definitely said which of these is the earlier. The difference in writing the symbols 1 to 3, seems to be due to the inherent difference between the two systems of writing. The principles upon which numerical signs are formed in the two systems are quite different.

Difference from other Notations. In the Brâhmî

[1] It has been incorrectly stated by Smith and Karpinski that the Nânâghât forms were vertical. See *Hindu Arabic Numerals*, p. 28.

there are separate signs for each of the numbers 1, 4 to 9 and 10, 20, 30, 40, 50, 60, 70, 80, 90, 100, 200, 300,...... 1000, 2000, etc., while in the oldest Kharoṣṭhī and in the earliest Semitic writings, the Hieroglyphic and the Phoenician, the only symbols are those for 1, 10, 20 and 100.

The Hieratic and the Demotic numerals, however, resemble the Brâhmî in having nineteen symbols for the numbers from 1 to 100, but the principle of formation of the numbers 200, 300, 400, 2,000, 3,000 and 4,000 are different, as will appear from Table II(*c*). The method of formation of intermediate and higher numbers is also different in the two systems. While the Brâhmî places the bigger numbers to the left, the arrangement is the reverse of this in the Kharoṣṭhī and Semitic writings. Thus the number 274, is written in Brâhmî with the help of the symbols for 200, 70 and 4 as (200) (70) (4), while in the Kharoṣṭhī and the Semitic numerals it is written as (4) (70) (200).[1]

Theories about their Origin. Quite a large number of theories have been advanced to explain the origin of the Brâhmî numerals. Points of resemblance have been imagined between these numerals and those of other nations. Recourse has been taken by writers to the turning, twisting, adding on or cutting off of parts of the numerals of other nations to fit their pet theories. It is needless to say that each of these theories had its own supporters who were quite convinced of the correctness of their explanations. We give below the outlines of some of these theories:

1. Cunningham[2] believed that writing had been known in India from the earliest known times, and

[1] Compare the same number written in Kharoṣṭhī, p. 24.

[2] Inscriptions of Aśoka, *Corpus Inscriptionum Indicarum*, Vol. I, p. 52.

that the earliest alphabet was pictographic. He suggested that the Brâhmî script was derived from the early pictographic writing. The theory is evidently capable of extension to the numerical signs. Later epigraphists, however, discarded the hypothesis as it appeared too fanciful to them. Cunningham's bold hypothesis regarding the antiquity of writing in India has been more than justified by the recent discovery of the use of a quasi-pictographic script on certain seals and in inscriptions belonging to the fourth millenium B.C. found amongst the excavations at Mohenjo-daro and Harappa. His theory has been revived by Langdon who is of opinion that the Brâhmî alphabet could be derived from the pictographs of Mohenjo-daro.[1] The theory is incomplete as the writings of Mohenjo-daro have not been completely deciphered as yet. It can be called a guess only. As regards the evolution of the Brâhmî numerals, it may be stated that it is at present extremely difficult to differentiate the numerical symbols from the Mohenjo-daro script. If the surmise that the figures, given on p. 19, are numerical symbols be correct, it will not be possible to develop a theory deriving the Brâhmî numerals from them.

2. Bayley[2] asserted that the principles of the Brâhmî system have been derived from the hieroglyphic notation of the Egyptians, and that the majority of the Indian symbols have been borrowed from Phoenician, Bactrian, and Akkadean figures or letters. As has been already remarked[3] the principles of the Brâhmî and the hieroglyphic systems are entirely different and

[1] *Mohenjo-daro etc.*, Chap. xii. This view is strongly supported by Hunter, *l.c.*, p. 490.

[2] *Journal of the Royal Asiatic Soc.*, XV, part I, reprint, London, 1882, pp. 12 and 17. The theory was supported by Taylor, *The Alphabet*, London, 1883, Vol. II, pp. 265-66.

[3] See pages 27-8.

unconnected. The reader will find the hieroglyphic and the Brâhmî systems shown together in Tables II (*a*), (*b*), (*c*), and convince himself of the incorrectness of Bayley's assertion. Moreover, the assumption that the Hindus borrowed from four or five different, partly very ancient and partly more modern, sources, is extremely difficult to believe. Regarding the resemblance between the Bactrian and Akkadean numbers and the Brâhmî forms postulated by Bayley, Bühler[1] remarks that in four cases (four, six, seven and ten) the facts are absolutely against Bayley's hypothesis. Some writers have also criticized Bayley's drawings as being affected by his theory.[2] Under these circumstances his derivation has to be rejected.

3. Burnell[3] pointed out the general agreement of the principles of the Indian system with those of the Demotic notation of the Egyptians. He asserted a resemblance between the Demotic signs for 1 to 9 and the corresponding Indian symbols, and put forward the theory that the Hindus borrowed these signs and later on modified them and converted them into *akṣaras* (letter forms).

4. Bühler[3] has put forward a modification of Burnell's theory. He states, "It seems to me probable that the *Brâhma* numerals are derived from the Egyptian Hieratic figures, and that the Hindus effected their transformation into *Akṣaras*, because they were already accustomed to express numerals by words."

The above theories like the one examined before are not well founded. Tables II (*a*), (*b*), (*c*), show the Hieratic and Demotic symbols together with those of the Brâhmî. An examination of the Tables will reveal

[1] Bühler, *On the Origin of the Indian Brâhma Alphabet*, Strassburg, 1898, pp. 52, 53 foot-note.
[2] *Cf.* Smith and Karpinski, *Hindu Arabic Numerals*, pp. 30-1.
[3] Bühler, *l.c.*, p. 82.

that out of the nineteen symbols to represent the numbers from 1 to 100, only the nine of the Brâhmî resembles the corresponding symbol of the Demotic or the Hieratic. There is absolutely no resemblance between any of the others. To base the derivation on a resemblance between the Hieratic 5 and the Brâhmî 7, as is sought to be done, is absurd. Likewise the changing and twisting of the Demotic and Hieratic forms to suit the theory is unacceptable.

That there is some resemblance between these systems in the fact that each employs the same number of signs, *i.e.*, nineteen, for the representation of numbers upto hundred, cannot be denied. There is, however, a difference in the method of formation of the hundreds and the thousands. In the Brâhmî the numbers 200 and 300 or 2,000 and 3,000, are formed by adding one *mâtṛkâ* and two *mâtṛkâs* to the right of the symbol for hundred or thousand respectively, thus

 = 100,  = 200,  = 300

 = 1,000,  = 2,000,  = 3,000.

The numbers 400 and 4,000 are formed by connecting the symbol for 100 and 1,000 to the number  (4), thus

 = 400 and  = 4,000.

In the Hieratic the corresponding symbols are:

 = 2

 = 3  = 100, = 200,

 = 1,000,  = 2,000,

 = 4,

 = 300,  = 3,000,  = 400, 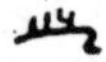 = 4,000.

It will be observed that in the Hieratic system the sign for one thousand is not used in the formation of the other thousands. The similarity in principle, even if it were complete, would not force us to conclude that one of these nations copied the other. The use of nineteen signs afforded the easiest and probably the best method of denoting numbers. It is not beyond the limits of probability that what appeared easy to the Egyptians might have also independently occurred to the Hindus.

There are on the other hand some considerations which make us suggest that the Egyptians borrowed the principles of the Hieratic and the Demotic systems from outside, and probably from India—a hypothesis which is not *a priori* impossible as it has been shown that the numeration system of the ancient Hindus based on nineteen signs might have been perfected about 1,000 B.C. It is known that the ancient Egyptian system employed only four signs, those for 1, 10, 20 and 100. Why should there be a sudden change from the old system to one containing nineteen signs cannot be adequately explained except on the hypothesis of foreign influence. Further, the cursive forms for the numbers 2, 3 and 4 are unsuited to the right to left Hieratic or Demotic script. Although these figures are connected with the earlier hieroglyphic and Phoenician figures, yet it is possible that the cursive combinations might have been formed to obtain the nineteen signs necessary for the new system, under the influence of a people with a left to right script. It may be, however, asserted that the hypothesis of an Indian origin of the Hieratic system is a mere suggestion. The two points noted above, by themselves, would not be enough, unless backed by other facts, to put forward a theory. It is expected that further discoveries will throw light on this point.

Relation with Letter Forms. It was suggested by James Princep,[1] as early as 1838, that the numerals were formed after the initial letters of the number names. But knowing the pronunciation of the number names, we find this not to be the case. Other investigators have held that the numeral signs were formed after the letters in the order of the ancient alphabet. Although we find that letters were used to denote numbers as early as the 8th century B.C.,[2] and that many systems of letter-numerals were invented in later times[3] and came into common use, yet we are forced to reject this hypothesis as resemblance between the old numerical forms and the letters in the alphabetic order cannot be shown to exist.

A peculiar numerical notation, using distinct letters or syllables of the alphabet, is found to have been used in the pagination of old manuscripts as well as in some coins and a few inscriptions. The signs are, however, not always the same. Very frequently they are slightly differentiated, probably in order to distinguish the signs with numerical values from those with letter values. The fact that these symbols are letters is also acknowledged by the name *akṣarapallî* which the Jainas occasionally give to this system, in order to distinguish it from the decimal notation, the *aṅkapallî*.[4]

[1] "Examination of inscriptions from Girnar in Gujerat, and Dhauli in Cuttack," *JASB*, 1838.

[2] The method seems to have been used by Pâṇini. See p. 63.

[3] *Vide infra*, pp. 64ff.

[4] Bühler, *l.c.*, p. 78. The details of the *akṣarapallî* are given later on (pp. 72ff).

The use of the letter system, the *akṣarapallî*, suggests that the old Brâhmî numerals also might have been fashioned after the letters or the syllables of the Brâhmî alphabet.

A detailed examination of the numerals and the letter-forms has been made, and the result is tabulated below:[1]

The numerals 1, 2 and 3 which were represented by horizontal strokes do not resemble any letter. They may have been derived from the *mâtṛkâ* signs.

4. The earliest form of this numeral in the inscriptions of Aśoka resembles the letter *ka*; the Nânâghât and Nâsik cave forms resemble the syllable *pka*. While this seems to have been the case with most of the later inscriptions, forms resembling the syllables *pna*, *lka*, *ṭka*, *pkṛ* are also to be met with.

5. In most of the inscriptions it resembles the syllable *tṛ*, whilst forms resembling *ta*, *tâ*, *pu*, *hu*, *ru*, *tṛ*, *tṛâ*, *nâ*, *na*, *hṛ*, *hra*, and *ha* are also found.

6. In most of the inscriptions it resembles *phra*, and in a few *phrâ*, *pha*, *phâ*, *ja* and *hâ*. The resemblance in this case is, however, not close enough.

7. In most of the inscriptions it resembles *gra* or *gu*, whilst n some it is like *ga*.

8. In most cases it resembles the syllable *hra* or *hrâ*. In some inscriptions the form of this numeral cannot be said to resemble any letter or syllable.

9. The oldest forms of this numeral, those in the Nânâghât, the Kuṣâna and the Kṣatrapa inscriptions, cannot be said to resemble any letter or syllable. The later forms, however, resemble the letter *u* or *om*.

[1] For the varying forms of the numerals see Tables III—XIII.

10. The oldest forms of this too do not resemble any letter or syllable, although later forms may be said to resemble *rya*, *ha*, *hṛ*, *kha*, and *ṭhu* or *ṭha*.[1]

20. In all cases it resembles the letter *tha*.
30. Resembles the letter *la* in all cases.
40. Resembles *pta* or *sa* in all cases.
50. It may be said to resemble the *anunâsika*.
60. Resembles *pu*, *pa* and *pra*.
70. Resembles *pu*, *pta*, *pra*, *pna*, *pnâ*, or *hra*.
80. Resembles the *upadhmânîya* sign.
90. Resembles the *upadhmânîya* sign with a central cross.
100. Resembles the syllable *su* in most of the inscriptions. In some it resembles *a*.

Indraji's Theory. The above details show that,

1. No phonetical value can be assigned to the ancient forms of the numerals 6 and 10, and 1, 2 and 3;

2. There is a great deal of variation in the phonetical values of the other units, excepting the cases of 7 and 9; and

3. The phonetical values of the tens are determinate excepting that of 70 which offers great variation.

Thus leaving aside the symbols for 1, 2 and 3, we find that out of the 16 symbols, no resemblance to letter-forms can be satisfactorily shown in the case of two (*i.e.*, 6 and 10),[2] and that in the case of three others (*i.e.*, 7, 9 and 70) there are too many variations in the phonetical values, whilst in the case of the remaining 11 symbols, the phonetical values are more or less determinate. These facts led Pandit Bhagavanlal Indraji

[1] According to Bühler, *l.c.*, p. 80.

[2] The resemblance to *ja*, *sa*, *phra*, etc. and *ṭhû*, *ṭha*, etc., respectively, stated by Bühler (*l. c.*, p. 80), does not appear to us to be close enough.

to put forward the hypothesis that the Brâhmî numerals are derived from the letters or syllables of the Brâhmî script. The Pandit, however, admitted his inability to find the key to the system, nor has it been found by any other scholar upto this time. The problem, in fact, appears to be insoluble, unless further epigraphic material is discovered to show the forms of the numerical symbols anterior to Aśoka. The Aśokan forms as well as those of later inscriptions are in a too well developed state, and are too far away from the time of invention of those symbols, to give us the desired information regarding their origin.

But of all the theories that have been advanced from time to time, that of Pandit Indraji seems to us to be the most plausible. The Hindus knew the art of writing in the fourth millennium B.C. They used numbers as large as 10^9 about 2,000 B.C., and since then their religion and their sciences have necessitated the use of large numbers. Buddha in the sixth century B.C. is stated to have given number names as large as 10^{53} and this number series was continued still further in later times.[1] All these facts reveal a condition that would have been impossible unless arithmetic had attained a considerable degree of progress. It is certain that the Hindus must have felt the necessity of some method of writing these numbers from the earliest known times. It would not be, therefore, against historical testimony to conclude that the Hindus invented the Brâhmî number system. The conclusion is supported by the use, in writing numbers, of the *mâtṛkâ*, the *anunâsika* and the *upadhmânîya* signs which are found only in the Sanskrit script and in no other script, whether ancient or modern. It is further strengthened by Indian tradition, Hindu, Jaina as well as Buddhist, which

[1] Cf. pp. 10-12.

ascribes the invention of the Brâhmî script and the numeral notation to Brahmâ, the Creator, and thereby claims it as a national invention of the remotest antiquity.[1]

Period of Invention. The invention of the system may be assigned to the period 1,000 B.C. to 600 B.C. As the Aśokan numerical figures indicate that the system was common all over India,[2] and that it has had a long history, the lower limit 1,000 B.C. is certainly not placed too early. On the other hand general considerations, such as the high development of the arts and the sciences, the mention of numerical signs and of 64 different scripts in ancient Buddhist literature,[3] and the use of large numbers at a very early period, all point to the date of the invention of the system as being nearer to 1,000 B.C., if not earlier.

Resume. The strength of Pandit Indraji's hypothesis lies in the fact that out of the nineteen signs, eleven definitely resemble the letters or the signs of the Brâhmî alphabet. The resemblance is too striking to be entirely accidental. Moreover, it has been found that the numerical forms closely followed the changing forms of the letters from century to century. This is especially true in the case of the tens and shows that the writers of the ancient inscriptions knew the phonetical values of these symbols. The divergence from letter forms in the case of the signs for the units may be due to the

[1] Bühler (*l.c.*, p. 1, foot-note,) quotes several authorities. Of these the *Nârada Smṛti* and the Jaina canonical work, the *Samavâyânga-sûtra*, belong to the fourth century B.C.

[2] Megasthenes speaks of mile-stones indicating distances and the halting places on the roads. *Indika of Megasthenes*, pp. 125-126; Bühler, *l.c.*

[3] Related in the *Lalitavistara*, both in the Sanskrit text and the Chinese translation of 308 A.D. The Jaina *Samavâyânga-sûtra* (*c.* 300 B.C.) and *Pannavanâ-sûtra* (*c.* 168 B.C.) each gives a list of 18 scripts; see Weber, *Indische Studien*, 16, 280, 399.

fact that they were the first to be invented and were in more common use, so that they acquired special cursive forms and did not follow the changes in the forms of the corresponding letters. We may now summarize the discussion given in this section by saying that (1) the Brâhmî numerical forms were undoubtedly of Indian origin, (2) the form of the tens were derived from certain letters or signs of the alphabet, and (3) the origin of the forms of the units is doubtful. It is probable that they, too, were fashioned after the letters of the alphabet, but there appears to be no means of justifying this assertion unless the forms of these numerals anterior to Aśoka are discovered.

8. THE DECIMAL PLACE-VALUE SYSTEM

Important Features. The third and most important of the Hindu numeral notations is the decimal place-value notation. In this system there are only ten symbols, those called *aṅka* (literally meaning "mark") for the numbers one to nine, and the zero symbol, ordinarily called *śûnya* (literally, "empty"). With the application of the principle of place-value these are quite sufficient for the writing of all numbers in as simple a way as possible. The scale is, of course, decimal. This system is now commonly used throughout the civilised world. Without the zero and the place-value, the Hindu numerals would have been no better than many others of the same kind, and would not have been adopted by all the civilised peoples of the world. "The importance of the creation of the zero mark," says Professor Halsted, "can never be exaggerated. This giving to airy nothing, not merely a local habitation and a name, a picture, a symbol, but helpful power, is the characteristic of the Hindu race whence it sprang.

It is like coining the *Nirvāṇa* into dynamos. No single mathematical creation has been more potent for the general on-go of intelligence and power."[1]

Forms. A large number of scripts differing from each other are in use in different parts of India today. The forms of the numerical signs in these scripts are also different. Although all the Hindu scripts are derived from a common source—the Brâhmî Script—yet the differences in the forms of the various modern Indian scripts are so great that it would have been difficult to establish any relation between them, if their previous history had not been known. The above remark applies to the numerical signs also, as will appear from a study of the numerical signs in the various vernaculars of India given in Table XV. The great divergence in the forms of the numerical symbols shows that in India, people already knew the use of the zero and the place-value principle before the different scripts came into being, and that the numeral forms were independently modified in various parts of India, just as the letters of the alphabet were modified. And as the changes in the forms in different localities were independent of each other, so there has come about a great divergence in the modern forms. That this divergence already existed in the eleventh century is testified to by Al-Bîrûnî who says, "As in different parts of India, the letters have different shapes the numerical signs, too, which are called *aṅka*, differ."[2]

Nagari Forms. The most important as well as the most widely used of the different symbols are those belonging to the *Nâgarî* script. The present forms of these symbols are:

[1] G. B. Halsted, *On the foundation and technique of Arithmetic*, Chicago, 1912, p. 20.

[2] *Alberuni's India*, I, p. 74.

१, २, ३, ४, ५, ६, ७, ८, ९, ०.

The gradual development of these figures from the Brâhmî numerals is shown in Table XIV.

Epigraphic Instances. The following is a list of inscriptions and grant plates upto the middle of the tenth century, which contain numerals written in the decimal place-value notation. The numerals in the inscriptions and plates after this period, are always given in decimal figures.

1. 595 A.D. Gurjara grant plate from Sankheda, (*EI*, II, p. 19). The date Saṁvat 346 is given in the decimal place-value notation.

*2. 646 A.D. Belhari Inscription, (*JA*, 1863).

*3. 674 A.D. Kanheri Inscription, (*JA*, 1863, p. 392).

4. 8th Century Ragholi plates of Jaivardhaṇa II, (*EI*, IX, p. 41). The number 30 is written in decimal figures.

5. 725 A.D. Two Sanskrit Inscriptions in the British Museum, (*IA*, XIII, p. 250). The dates Saṁvat 781 (=723 A.D.) and Saṁvat 783 (=725 A.D.) are given in decimal figures.

*6. 736 A.D. Dhiniki copper plate grant, (*IA*, XII, p. 155). The date Vikrama Saṁvat 794 is given in decimal figures.

7. 753 A.D. Ciacole plates of Devendravarmaṇa, (*EI*, III, p. 133). The number 20 is written in decimal figures.

8. 754 A.D. Râṣṭrakûṭa grant of Dantidurga, (*IA*, XI, p. 108). The date Saṁvat 675 is given in decimal figures.

9. 791 A.D. Inscription of Sâmanta Devadatta, (*IA*, XIV, p. 351). The date Vikrama Saṁvat 847 is given in decimal figures.

10. 793 A.D. Daulatabad plates of Śaṅkargaṇa, (*EI*, IX, p. 197). The date Śaka 715 is given in decimal figures.

*11. 813 A.D. Torkhede plates, (*EI*, III, p. 53; also *IA*, XXV, p. 345). The date Śaka Saṁvat 735 is given in decimal figures.

12. 815 A.D. Buchkalâ inscription of Nâgbhaṭa, (*EI*, IX, p. 198). The date Saṁvat 872 is given in decimal figures.

13. 837 A.D. Inscription of Bâuka (Rajputana Museum, *PLM*, p. 127; *EI*, XVIII, p. 87). The date Vikrama Saṁvat 894 is given in decimal figures.

14. 843 A.D. The inscriptions from Kanheri, No. 43 b., (*IA*, VIII, p. 133). The date Saṁvat 765 is given in decimal figures.

15. 851 A.D. The inscriptions from Kanheri, No. 15, (*Ibid*). The date[1] Saṁvat 775 is given in decimal figures.

16. 853 A.D. Pâṇḍukeśvara Plates of Lalitasuradeva, (*IA*, XXV, p. 177). The date Saṁvat 21 of the King's reign is given in decimal figures.

17. 860 A.D. Ghatiyala Inscription of Kakkuka (*EI*, IX, p. 277). The date Vikrama Saṁvat 918 is given in decimal figures.

[1] For correction of date see *IA*, XX, p. 421.

18. 862 A.D. Deogarh Jaina Inscription of Bhojadeva, (*EI*, IV, p. 309). The dates Vikrama Saṁvat 919 and the corresponding Śaka Saṁvat 784 are both given in decimal figures.

19. 870 A.D. Gwalior inscription of the reign of Bhojadeva (*Archaeological Survey of India*, Report, 1903-4, plate 72). Although the date is not given, the ślokas are numbered from 1 to 26 in decimal figures.

20. 876 A.D. Gwalior inscription of Allah, of the reign of Bhojadeva (*EI*, I, p. 159). The date Vikrama Saṁvat 933, as well as the numbers 270, 187 and 50 are given in decimal figures.

21. 877 A.D. The inscriptions from Kanheri, No. 43a, (*IA*, XIII, p. 133). The date Saṁvat 799 is given in decimal figures.

22. 882 A.D. Pehava inscription (*EI*, I, p. 186). The date Saṁvat 276 (Śrî Harṣa Era) is given in decimal figures.

23. 893 A.D. Grant plate of Balavarmaṇa, (*EI*, IX, p. 1). The date Vallabhî Saṁvat 574 is given in decimal figures.

24. 899 A.D. Grant plate of Avanîvarmaṇa, (*EI*, IX, p. 1). The date Vikrama Saṁvat 956 is given in decimal figures.

25. 905 A.D. The Ahar stone inscription (*Journ. United Provinces Hist. Soc.*, 1926, pp. 83 ff) contains several dates written in decimal figures.

26. 910 A.D. Râṣṭrakûṭa grant of Krishna II (*EI*,

		I, p. 53). The date is given in decimal figures.
27.	917 A.D.	Sanskrit and old Canarese inscriptions, No. 170, (*IA*, XVI, p. 174). The date Saṁvat 974 is given in decimal figures. The number 500 also occurs.
28.	930 A.D.	Cambay plates of Govinda IV, (*EI*, VII, p. 26). The date Śaka Saṁvat 852 is given in decimal figures.
29.	933 A.D.	Sangli plates of Râṣṭrakûṭa Govindarâja IV, (*IA*, XII, p. 249). The date Saṁvat 855 is given in decimal figures.
30.	951 A.D.	Sanskrit and old Canarese inscriptions, No. 135, (*IA*, XII, p. 257). The date Saṁvat 873 is given in decimal figures.
31.	953 A.D.	Inscription of Yaśovarmaṇa, (*EI*, I, p. 122). The date Saṁvat 1011 is given in decimal figures.
32.	968 A.D.	Siyadoni stone inscription (*EI*, I, p. 162). The inscription contains a large number of numerals expressed in decimal figures.
33.	972 A.D.	Râṣṭrakûṭa grant of Amoghavarṣa, (*IA*, XII, p. 263). The date Śaka 894 is given in decimal figures.

Palaeographic evidences of the early use of the decimal place-value system of notation are found in the Hindu colonies of the Far East.[1] The most important ones among these are the three inscriptions of

[1] G. Coedès, "A propos de l'origine des chiffres arabes," *Bull. School of Oriental Studies* (London), VI, 1931, pp. 323-8.

Śrîvijaya, two found at Palembang in Sumatra, and the third in the island of Banka. These contain the dates 605, 606 and 608 of the Śaka Era (corresponding respectively to A.D. 683, 684 and 686) written in numerical figures. Another instance giving the date 605 Śaka is the inscription of Sambor in Cambodia. In an inscription at Po Nagar in Champa, occurs the date 735 Śaka (= 813 A.D.).

Their Supposed Unreliability. The above list contains more than thirty undoubted epigraphic instances of the use of the place-value notation in India. G.R.Kaye,[1] who believes in the theory of the non-Hindu origin of the place-value notation, states that all the early epigraphic evidences of its use in India are unreliable. On the basis of the existence of a few forged grant plates he asserts that in the eleventh century "there occurred a specially great opportunity to regain confiscated endowments and to acquire fresh ones" and thereby concludes that all early epigraphic evidences must be unreliable. Such reasoning is obviously fallacious and needs no refutation.

Most of the copper plates are legal documents recording gifts made by rich persons or kings to Brâhmaṇas on religious occasions. The plates contain details as to the occasion for making the gifts, the names of the donor and the donee, the description of the movable and immovable properties transferred by the gift, and the date of the gift which is always written out in full in words and very often in figures also. The forgeries may be of two kinds: (1) In the original documents, parts relating to either the names of the donor or the donee, or the description of the immovable property may have been obliterated by being beaten out and new

[1] "Notes on Indian Mathematics," *JASB*, (N. S., 1907), III, pp. 482-87.

names or descriptions substituted. All such forgeries are easily detected, because of the uneven surface of the part of the plate that is tampered with and the difference in the writing. (2) An entirely new document may be forged. Such cases, though rare, are also easily detected, because there is obvious divergence as to the date recorded in the document, and that inferred on the basis of the forms of the characters used in the writing. Such forgeries are also marked by an obvious inferiority in execution, and inaccuracies in the statement of genealogies and other historical facts.

Epigraphists have so far found little difficulty in eliminating the spurious grant plates. It might be mentioned that the genuineness of the grant plates included in our list has not been questioned by any epigraphist.[1]

Kaye, in his article quoted above, has given a list of eighteen inscriptions and grant plates and eliminates all but the last two as forgeries. The arguments he has employed and the assertions of facts that he has made are in most cases incorrect and misleading, so that his conclusions cannot be accepted. As an instance of his method, we quote his criticism about the Gurjara grant plate, No. 1 in our list. He writes: "Dr. Bühler quotes this Gurjara inscription of the Chedi year 346 or A.D. 594 as the earliest epigraphic instance of the use of the decimal notation in India. (i) An examination

[1] If any of them is forged, the forgery is so good that it cannot be detected. The writing in such cases, if any, is so well forged as to be indistinguishable from that used in the period to which the plate is said to belong. Therefore, the evidence of these plates as to the method of writing numbers, cannot be rejected, even if they be proved to be spurious at some future date—a contingency which is very unlikely to use. It may also be noted that the list contains several *stone inscriptions* which cannot be spurious.

of the plate (*Ep. Ind.*, II., p. 20) suggests the possibility that the figures were added some time after the plate was engraved. The date is engraved in words as well as in figures. It is 'three hundred years exceeded by forty-six.' The symbols are right at the end of the inscription from which they are marked off by a double bar in a most unusual manner. (ii) The figures are of the type of the period, but they were also in use much later, and in no other example are such symbols used with place-value. (iii) Also there are nine dates written in the old notation (*Ep. Ind.*, V), *e. g.*, there is another grant of the Gurjara of Bharoch in which the date Saṁvat 391 (i.e., A.D. 640) is given in the old notation. Again, there is no other Chedi date, at least before the eleventh century A.D., given in the modern (place-value) notation. (iv) There cannot be the remotest doubt as to the unsoundness of this particular piece of evidence of the early use of the modern system of notation in India."

The following remarks will show to the reader that Kaye's criticism and his conclusion are unfounded and invalid:

(*i*) An examination of the plate, (*EI*, II, p. 19), will convince every one about its genuineness. The writing is bold and clear, the numerical figures occur at the end, as they ought to be, immediately after the words 'three hundred years exceeded by forty-six.' They are separated from the written words by bars, just as they ought to be. There is absolutely nothing suspicious about this method of separation, as it is common custom in India to do so and occurs frequently. That it was the practice to write the date at the end of a document is well known.[1] In fact, the numeral

[1] Many of the plates mentioned in our list contain the date at the end.

figures of the date occasionally mark the end of the document.[1] The double vertical bar, ||, is a sign of inter-punctuation. Although punctuation marks have been in use in India from the earliest known times, yet their use did not become either regular or compulsory till very recent times.[2] Different writers used the various marks differently. In inscriptions, the double vertical bar has been found at the end of sentences, half verses, verses, larger prose sections and documents. In the Junâr inscriptions it occurs after numerals and once after the name of the donor.[3] In manuscripts, the practice of separating numbers by vertical bars is common. It is found in the Bakhshâlî Manuscript[4] and in several others. Thus the occurrence of the numerals at the end and the inter-punctuation mark of the double vertical bar cannot form valid grounds for suspecting the document. The suggestion that the figures were added some time after the plate was engraved is absurd, as there appears to us no reason why one should take the trouble to add the figures when the date was already written in words.

(*ii*) Kaye admits that the figures are of the type of the period. His remark that they were in use much later is incorrect. The Tables III-V and XII show that the use of three horizontal bars to represent 3 is not

[1] This is so in the Chârgâon plates of Huviṣka (*Arch. Survey Report*, 1908-9, plate 56), in the Inscription of Rudradamana (*IA*, VIII, p. 42) and in others.

[2] There are some copper plate grants which do not contain any punctuation marks; see Bühler, *l.c.*, p. 90.

[3] Bühler, *l.c.*, p. 89.

[4] *E.g.*, | 5 |, 21 r; | 2558 |, 2v; | 330 |, 17v; instances such as these: | 1 | 4 | 9 | 16 |, 16v; and |2|, |4|, etc., 5v. are very common. Very often, isolated numbers are not separated. The double vertical bar also occurs before and after the words *udâ*, *sûtraṃ*, etc.

found after the eighth century. The figure for 4 used in our grant plate is not found after the sixth century, and the same is true for the figure for 6. The forms of the numerical signs alone fix the date of the writing to the sixth century and not later.

(*iii*) The *Chedi Saṁvat* is one of the thirty-four eras, whose use has been discovered in inscriptions and grant plates. The occurrence of nine dates in the Chedi Saṁvat, written in the old notation after this plate, does not prove the unsoundness of this particular piece of evidence, as Kaye would like us to conclude. It simply shows that in India too, the new system had to fight for supremacy over the older one just as in other countries. In Arabia the new system was introduced in the eighth century, but it did not come into common use until five or six hundred years later. In Europe we find that it was exceptional for common people to use the new system before the sixteenth century—a good witness to this fact being the popular almanacs. Calendars of 1557-96 have generally Roman numerals, while Koebel's Calendar of 1578 gives the Hindu numerals as subordinate to the Roman.[1]

We may, therefore, conclude that the Gurjara grant plate offers us a genuine instance of the use of the new system (with place-value) in India.

Kaye's criticisms regarding the genuineness of some other plates included in our list (marked with asterisks) have been found to be baseless.

Place of Invention of the New System. It has been already stated that the same numeral forms for the numbers 1 to 9, as were in use in India from the earliest known times, have been used in the new system of notation with the place-value. Another noteworthy fact

[1] Smith and Karpiński, *l.c.*, p. 133.

regarding the new system is the arrangement of the *aṅka* (digits). It will be observed that the arrangement in the old system was that the bigger numbers were written to the left of the smaller ones.[1] This same arrangement continues in the new system with place-value, where the digits to the left, due to their place or position, have bigger values. The gradual change from the old system to the new one using the same numerical signs, is to be found in India alone, and this, in our opinion, is one of the strongest arguments in favour of the Hindu origin of the new system. The earliest epigraphic instance of the use of the new system is 594 A.D. No other country in the world offers such an early instance of its use. Epigraphic evidence alone is, therefore, sufficient to assign a Hindu origin to the modern system of notation.

Inventor Unknown. It is not known who the inventor of the new system was, and whether it was invented by some great scholar, or by a conference of sages or by gradual development due to the use of some form of the abacus. Likewise, it is not known to which place, city, district or seat of learning belongs the honour of the invention and its first use. Epigraphic evidence cannot help us in this direction. For the system was used in inscriptions, a very long time after its invention, in fact, when it had become quite popular all over Northern India.

Time of Invention. The grant plates were legal documents. They were written by professional writers. The existence of such writers is mentioned in the southern Buddhist canons and in the Epics.[2] They have

[1] Showing thereby that the place assigned to a numeral depended upon its value. This has been incorrectly thought to be a sort of place-value system by some writers.

[2] Bühler, *l.c.*, p. 5.

been called *lekhaka*, *lipikara* and later on *divira*, *karaṇa*, *kâyastha*, etc. According to Kalhaṇa,[1] the Kings of Kashmir employed a special officer for drafting legal documents. He bore the title of *paṭṭopâdhyâya*, *i.e.*, the teacher (charged with the preparation) of title deeds. The existence of manuals such as the *Lekhapañcâśikâ*, the *Lekhaprakâśa*, which give rules for drafting letters, land grants, treaties, and various kinds of bonds and bills of exchange, show beyond doubt that the writing of grant plates was a specialised art and that the style of writing those documents must always have been centuries behind the times, just as it is even to-day with respect to legal and state documents. The time of invention of the new system must, therefore, be placed several centuries before its first occurrence in a grant plate in the sixth century A.D. The exact period of invention may be roughly deduced from the history of the growth of numerical notations in other countries.

According to Heath,[2] the Greek alphabetic notation was invented in the 7th century B.C., but it came into general use only in the second century A.D. Thus it took about eight hundred years to get popular. In Arabia the new notation was introduced in the 8th century A.D., but it came into common use about five or six hundred years later. The same was the case in Europe. The Arabs got the complete decimal arithmetic, including the method of performing the various operations, at a period when intellectual activity in Arabia was at its greatest height, but they could not make the decimal system common before about five or six hundred years had elapsed.[3] In legal documents

[1] *Râjataraṅgiṇî*, V, pp. 397f.

[2] Heath, *History of Greek Mathematics*, I, Oxford, 1921, p. 34.

[3] The arithmetic written by Al-Kharki in the eleventh century does not use the decimal system, showing that at the time there were two schools amongst the Arab mathematicians, one favouring

and in recording historical dates, the Arabs even now use their old alphabetic notation.

Epigraphic evidences show that the new system was quite common in India in the eighth century and that the old system ceased to exist in Northern India by the middle of the tenth century. This would, therefore, place the invention of our system in the period between the first century B.C. and the third century A.D.

The exact date of the invention, however, would be nearer to the 1st century B.C. or even earlier, because for a long time after its invention, the system must have been looked upon as a mere curiosity and used simply for expressing large numbers. A still longer time must have elapsed before the method of performing the operations of addition, subtraction, multiplication, division and the extraction of roots, could be perfected. It would be only after the perfection of the methods of performing the operations that the system could be used by mathematicians. And then after this it would take about five hundred years, as in Arabia, to become popular. There should, therefore, be a gap of about eight centuries between the time of invention and its coming into popular use, just as was the case with the Greek alphabetic notation. Therefore, on epigraphic evidence alone, the invention of the place-value system must be assigned to the beginning of the Christian era, very probably the 1st century B.C. This conclusion is supported by literary and other evidences which will be given hereafter.

the Hindu numerals, while the other stuck to the old notation. See the article on "Hisâb" by H. Suter in the *Encyclopaedia of Islam*.

9. PERSISTENCE OF THE OLD SYSTEM

The occurrence of the old system of writing numbers, with no place-value, is found generally in inscriptions upto the seventh century A.D., after which it was gradually given up in favour of the new system with place-value. Occasional use of the old system, however, is to be met with in Nepal and in some South Indian inscriptions upto the beginning of the tenth century A.D., but after this period the old system seems to have been forgotten, and completely gone out of use. In the seventh century the new system was in general use, but the old system seems to have been given preference in inscriptions. There are a number of grant plates of the eighth century A.D., in which the dates, although written in the old notation, are incorrectly inscribed, showing thereby that people had already forgotten the old system. In a grant plate of Sîlâditya VI,[1] dated the Gupta year 441 (*c.* 760 A.D.), the sign for 40, instead of the sign for 4, has been subjoined to the sign for 100 to denote 400, *i.e.*, 4,000 has been incorrectly written for 400. There is another grant plate, dated the Gâṅgeya year 183 (*c.* 753 A.D.), in which the figure 183 is wrongly written.[2] This plate is of special interest as it exhibits the use of the old and the new systems in the same document.[3] Another very interesting instance of the use of the old and the new systems in one and the same document is the Ahar stone

[1] *IA*, VI, p. 19, (plate).

[2] *EI*, III, p. 133, (plate). In this the sign of 8 is written for 80 and that of 30 for 3. The number 20 has been written by placing a dot after 2.

[3] For other instances showing admixture of both the old and the new systems, see Fleet Gupta Inscriptions, *Corpus Inscriptionum Indicarum*, III, p. 292; also *IA*, XIV, p. 351, where (800) (4) (9)=849.

inscription.[1] The document records gifts made on several occasions ranging over thirty-seven years, the last entry corresponding to 905 A.D. In this inscription the old notation is used in the first six lines whilst in the following lines it has been discarded and the new place-value notation appears. It is evident from the forms that the writer did not know the old system. For instance, 200 is written by adding the subscript 2 to the letter *su* (100), instead of using a *mâtṛkâ* sign as in the old system. In the same way the sign for 10 is incorrect in so far as a small zero has been affixed to the usual sign for ten. The inscription shows that although the old system had gone out of use completely, yet people tried to use it in inscriptions, probably for the same reason that makes us use the Roman numerals in giving dates, in numbering chapters of books, and in marking the hours on the face of a clock, even upto the present day.

10. WORD NUMERALS

Explanation of the System. A system of expressing numbers by means of words arranged as in the place-value notation was developed and perfected in India in the early centuries of the Christian era. In this system the numerals are expressed by names of things, beings or concepts, which, naturally or in accordance with the teaching of the *Śâstras*, connote numbers. Thus the number one may be denoted by anything that is markedly unique, *e.g.*, the moon, the earth, etc.; the number two may be denoted by any pair, *e.g.*, the eyes, the hands, the twins, etc.; and similarly others. The zero is denoted by words meaning void, sky, complete, etc.

[1] C. D. Chatterjee, "The Ahar stone inscription," *Journ. United Provinces Hist. Soc.*, 1926, pp. 83-119.

The system is used in works on astronomy, mathematics and metrics, as well as in the dates of inscriptions and in manuscripts. The ancient Hindu mathematicians and astronomers wrote their works in verse. Consequently they strongly felt the need for a convenient method of expressing the large numbers that occur so often in the astronomical works and in the statement of problems in mathematics. The word numerals were invented to fulfil this need and soon became very popular. They are used even upto the present day, whenever big numbers have to be expressed in Sanskrit verse.

The words denoting the numbers from one to nine and zero, with the use of the principle of place-value, give us a very convenient method of expressing numbers by word chronograms. To take a concrete case, the number 1,230 may be expressed in many ways:

1. *kha-guṇa-kara-âdi*,
2. *kha-loka-karṇa-candra*,
3. *âkâśa-kâla-netra-dharâ*, etc.

It will be observed that the same number can be expressed in hundreds of ways by word chronograms. This property makes the word numerals specially suitable for inclusion in metre. To secure still greater variety, the numbers beyond ten are also sometimes denoted by words.

List of Words. The following is a list of words commonly used in this system to denote numbers:

0 is expressed by *śûnya*, *kha*, *gagana*, *ambara*, *âkâśa*, *abhra*, *viyat*, *vyoma*, *antarikṣa*, *nabha*, *jaladharapatha*, *pûrṇa*, *randhra*, *viṣṇupada*, *ananta*, etc.

1 is expressed by *ádi*, *śaśi*, *indu*, *vidhu*, *candra*, *kalâdhara*, *himagu*, *śîtáṁśu*, *kṣapâkara*, *himâṁśu*, *sîtaraśmi*, *práleyáṁśu*, *soma*, *śaśâṅka*, *mṛgâṅka*, *himakara*, *sudhâṁśu*, *rajanikara*, *śaśadhara*, *śveta*, *abja*, *bhû*,

bhûmi, *kṣiti*, *dharâ*, *urvarâ*, *go*, *vasundharâ*, *pṛthvî*, *kṣmâ*, *dharaṇî*, *vasudhâ*, *ilâ*, *ku*, *mahî*, *rûpa*, *pitâmaha*, *nâyaka*, *tanu*, etc.

2 is expressed by *yama*, *yamala*, *aśvin*, *nâsatya*, *dasra*, *locana*, *netra*, *akṣi*, *dṛṣṭi*, *cakṣu*, *aṁbaka*, *nayana* *îkṣaṇa*, *pakṣa*, *bâhu*, *kara*, *karṇa*, *kuca*, *oṣṭha*, *gulpha*, *jânu*, *jaṅghâ*, *dvaya*, *dvanda*, *yugala*, *yugma*, *ayana*, *kuṭumba*, *ravicandrau*, *naya*,[1] etc.

3 is expressed by *râma*, *guṇa*, *triguṇa*, *loka*, *trijagat*, *bhuvana*, *kâla*, *trikâla*, *trigata*, *trinetra*, *haranetra*, *sahodarâḥ*, *agni*, *anala*, *vahni*, *pâvaka*, *vaiśvânara*, *dahana*, *tapana*, *hutâśana*, *jvalana*, *śikhin*, *kṛśânu*, *hotṛ*, *pura*, *ratna*[2] (Jaina), etc.

4 is expressed by *veda*, *śruti*, *samudra*, *sâgara*, *abdhi*, *ambhodha*, *ambhodhi*, *jaladhi*, *udadhi*, *jalanidhi*, *salilâkara*, *viṣanidhi*, *vâridhi*, *payodhi*, *payonidhi*, *ambudhi*, *kendra*, *varṇa*, *âśrama*, *yuga*, *turya*, *kṛta*, *aya*, *âya*, *diṣ*, *bandhu*, *koṣṭha*, *gati*, *kaṣâya*, etc.

5 is expressed by *bâṇa*, *śara*, *śastra*, *sâyaka*, *iṣu*, *bhûta*, *parva*, *prâṇa*, *pavana*,[3] *pâṇḍava*, *artha*, *viṣaya*, *mahâbhûta*, *tatva*, *bhâva*, *indriya*, *ratna*, *karaṇîya*,[4] *vrata*, etc.

6 is expressed by *rasa*, *aṅga*, *kâya*, *ṛtu*, *mâsârdha*, *darśana*, *râga*, *ari*, *śâstra*, *tarka*, *kâraka*, *lekhya*, *dravya*,[5] *khara*, *kumâravadana*, *śaṇmukha*, etc.

7 is expressed by *naga*, *aga*, *bhûbhṛt*, *parvata*, *śaila*, *acala*, *adri*, *giri*, *ṛṣi*, *muni*, *yati*, *atri*, *vâra*, *svara*,

[1] Method of comprehending things from particular standpoints—*dravyârthika* and *paryâyârthika*.

[2] Used by Mahâvîra only; others take it for five.

[3] See *SiŚe*, i. 27; *SiŚi*, *gaṇitâdhyâya*, x. 2. Used also for 7 (See the quotations by Bhaṭṭotpala in his commentary on *Bṛhatsaṁhitā*, ch. ii). In Al-Bîrûnî's list it is erroneously put for 9.

[4] That which ought to be done; according to the Jainas—*ahiṁsâ*, *sunṛta*, *asteya*, *brahmacarya*, and *aparigraha*.

[5] Used by Mahâvîra.

dhâtu, *aśva*, *turaga*, *vâji*, *haya*, *chandaḥ*, *dhî*, *kalatra*, *tatva*,[1] *dvîpa*, *pannaga*,[2] *bhaya*,[3] *mâtṛkâ*, *vyasana*, etc.

8 is expressed by *vasu*, *ahi*, *nâga*, *gaja*, *danti*, *dvirada*, *diggaja*, *hastin*, *ibha*, *mâtaṅga*, *kuñjara*, *dvîpa*, *puṣkarin*, *sindhura*, *sarpa*, *takṣa*, *siddhi*, *bhûti*, *anuṣṭubha*, *maṅgala*, *anîka*, *karman*,[4] *durita*, *tanu*,[5] *dik*,[6] *mada*,[7] etc.

9 is expressed by *aṅka*, *nanda*, *nidhi*, *graha*, *randhra*, *chidra*, *dvâra*, *go*,[8] *upendra*, *keśava*, *târkṣyadhvaj*, *durgâ*, *padârtha*,[9] *labdha*, *labdhi*, etc.

10 is expressed by *diś*, *dik*, *diśâ*, *âśâ*, *aṅgulî*, *paṅkti*, *kakubh*, *râvaṇaśira*, *avatâra*, *karman*, etc.

11 is expressed by *rudra*, *îśvara*, *mṛḍa*, *hara*, *îśa*, *bhava*, *bharga*, *śulin*, *mahâdeva*, *akṣauhiṇî*, etc.

12 is expressed by *ravi*, *sûrya*, *ina*, *arka*, *mârtaṇḍa*, *dyumaṇi*, *bhânu*, *âditya*, *divâkara*, *mâsa*, *râśi*, *vyaya*, etc.

13 is expressed by *viśvedevâḥ*, *viśva*, *kâma*, *atijagatî*, *aghoṣa*, etc.

14 is expressed by *manu*, *vidyâ*, *indra*, *śakra*, *loka*,[10] etc.

15 is expressed by *tithi*, *ghasra*, *dina*, *ahan*, *pakṣa*,[11] etc.

16 is expressed by *nṛpa*, *bhûpa*, *bhûpati*, *aṣṭi*, *kalâ*, etc.

17 is expressed by *atyaṣṭi*, etc.

[1] Used by Mahâvîra because the Jainas recognise seven *tatvas*; used for five by others.

[2] Used by Mahâvîra.

[3] Used by Mahâvîra.

[4] Used by Mahârvîra for 8 and by others for 10.

[5] Used by Mahâvîra.

[6] This word has been used for 8 as well as for 10. The use of *diś* or *dik* for 4 also occurs.

[7] Used by Mahâvîra only.

[8] This has been used for 1 also.

[9] Used by Mahâvîra only.

[10] Also used for 3.

[11] Also used for 2.

18 is expressed by *dhṛti*, etc.
19 is expressed by *atidhṛti*, etc.
20 is expressed by *nakha*, *kṛti*, etc.
21 is expressed by *utkṛti*, *prakṛti*, *svarga*, etc.
22 is expressed by *kṛti*, *jâti* (?), etc.
23 is expressed by *vikṛti*.
24 is expressed by *gâyatrî*, *jina*, *arhat*, *siddha*, etc.
25 is expressed by *tatva*,[1] etc.
27 is expressed by *nakṣatra*, *uḍu*, *bha*, etc.
32 is expressed by *danta*, *rada*, etc.
33 is expressed by *deva*, *amara*, *tridaśa*, *sura*, etc.
48 is expressed by *jagatî*, etc.
49 is expressed by *tâna*, etc.

Word Numerals without Place-value. In the *Veda* we do not find the use of names of things to denote numbers, but we do find instances of numbers denoting things. For instance, in the *Ṛgveda* the number 'twelve' has been used to denote a year[2] and in the *Atharvaveda* the number 'seven' has been used to denote a group of seven things (the seven seas, etc.).[3] There are instances, however, of fractions having been denoted by word symbols, *e.g.*, *kalâ* = $\frac{1}{16}$, *kuṣṭha* = $\frac{1}{12}$, *śapha* = $\frac{1}{4}$.

The earliest instances of a word being used to denote a whole number are found about 2,000 B.C., in the *Śatapatha Brâhmaṇa*[4] and *Taittirîya Brâhmaṇa*.[5] The

[1] Generally used for 5; also for 7 by Mahâvîra.

[2] "*Deva hitiṃ jugupurdvâdaśasya ṛtuṃ narona praminantyete*. . . . . ." (vii. 103, 1).

[3] "*Om ye triśapta pariyante*. . . ." (i. 1, 1).

[4] The word *kṛta* has been used for 4. "*catuṣṭomena kṛtena ayânâṃ*. . . ." (xiii. 3. 2. 1).

[5] "*Ye vai catvâraḥ stomâḥ kṛtaṃ tat*. . . ." (i. 5. 11. 1).

Chândogya Upaniṣad also contains several instances. In the *Vedâṅga Jyotiṣa*[1] (1,200 B.C.) words for numerals have been used at several places. The *Srauta-sûtras* of *Kâtyâyana*[2] and *Lâtyâyana*[3] have the words *gâyatrî* for 24 and *jagatî* for 48.

At this early stage, however, the word symbols were nothing more than curiosities; their use to denote numbers was rare. Moreover, we find evidences of a certain indefiniteness in the numerical significance attached to certain words. For instance, in the same work, the *Aitareya Brâhmaṇa*, the word *virâṭ* has been used to denote 10 at one place and 30 at another. The principle of place-value being unknown, the word symbols could not be used to denote large numbers, which were usually denoted in terms of the numerical denominations or by breaking the number into parts.[4] The use of the word symbols without place-value is found in the *Piṅgala Chandaḥ-sûtra* composed before 200 B.C. The principle of place-value seems to have been applied to the word numerals between 200 B.C. and 300 A.D.

Word Numerals with Place-value. The earliest instance of the use of the word numerals with place-value in its current form is found in the *Agni-Purâṇa*,[5]

[1] *rûpa* = 1, *aya* = 4, *guṇa* = *yuga* = 12, *bhasamûha* = 27. See (YJ. 23, ÂJ. 31), (YJ. 13, ÂJ. 4), (ÂJ. 19), (YJ. 25) and (YJ. 20) respectively.

[2] Weber's edition of *Kâtyâyana Śrauta Sûtra*, p. 1015.

[3] ix. 4. 31.

[4] E.g.,

Daśâyutânâmayutaṁ sahasrâṇi ca viṁśatiḥ
Koṭyaḥ ṣaṣṭiśca ṣaṭ caiva yo'smin rajan-mṛidhe hatâḥ

that is, 10 (10000) + 10000+20 (1000) + 60 (10,000,000) + 6 (10,000,000)—*Mahâbhârata*, *Strîparva*, xxvi. 9.

[5] *Agni-Purâṇa*, Baṅgabâsî ed., Calcutta (1314 B.S.), chs. 122-23, 131, 140, 141, 328-335. According to Pargiter, probably the greatest Puranic scholar of modern times, "the *purâṇas* cannot

a work which belongs to the earliest centuries of the Christian era. Bhaṭṭotpala in his commentary on the *Bṛhat-saṁhitâ* has given a quotation from the original *Puliśa-siddhânta*[1] (*c*. 400) in which the word system is used. The number expressed in this quotation is *kha* (0) *kha* (0) *aṣṭa* (8) *muni* (7) *râma* (3) *aśvi* (2) *netra* (2) *aṣṭa* (8) *śara* (5) *râtripâḥ* (1) = 1,582,237,800. There are in this work[2] several other quotations from the *Puliśa-siddhânta*, which contain word numerals. Later astronomical and mathematical manuals such as the *Sûrya-siddhânta* (*c*. 300), the *Pañca-siddhântikâ*[3] (505), the *Mahâ-* and *Laghu-Bhâskarîya*[4] (522), the *Brâhma-sphuṭa-siddhânta*[5] (628), the *Triśatikâ*[6] (*c*. 750), and the *Gaṇita-sâra-saṁgraha*[7] (850), all make use of the word notation.[8]

Word Numerals in Inscriptions. The earliest epigraphic instances of the use of the word numerals are met with in two Sanskrit inscriptions[9] found in Cambodia which was a Hindu colony. They are dated 604

be later than the earliest centuries of the Christian era." (*JRAS*, 1912, pp. 254-55). The *Agni-Purâṇa* is admitted by all scholars to be the earliest of the *Purâṇas*.

[1] *Bṛhat-saṁhitâ*, ed. by S. Dvivedi, Benares, p. 163.

[2] *Ibid*, pages 27, 29, 49, 51, etc. We are, however, not sure whether those quotations are from the original work or from a later redaction of the same.

[3] i. 8; viii. 1, etc.

[4] See *MBh*, ch. 7 and *LBh*, ch. 1.

[5] i. 51-55, etc.

[6] R. 6, Ex. 6, etc.

[7] ii. 7, 9, etc.

[8] In the face of the evidence adduced here, G. R. Kaye's assertion, (*Indian Mathematics*, Calcutta, 1915, p. 31) that the word numerals were introduced into India in the ninth century from the east, shows his ignorance of Indian mathematical works, or is a deliberate misrepresentation.

[9] R. C. Mazumdar, *Ancient Indian colonies in the far east,—Campa*, Vol. I, Lahore, 1927; see inscriptions Nos. 32, 39; also 40, 41, 43 and 44.

A.D. and 625 A.D. Their next occurrence is found in a Sanskrit inscription of Java, belonging to the 8th century.[1]

In India proper, although they were in use amongst the astronomers and mathematicians from the 3rd or 4th century A.D. onwards, it did not become the fashion to use them in inscriptions till a much later date. The earliest Hindu inscriptions using these numerals are dated 813 A.D.[2] and 842 A.D.[3] In the following century they are used in the plates issued by the Eastern Chalukya Amma II, in 943 A.D.[4] In later times the epigraphic instances become more frequent. The notation is also found in several manuscripts in which dates are given in verse.[5]

Origin and Early History. It should be noted that the arrangement of words, representing the numbers zero and one to nine, in a word chronogram is contrary to the arrangement that is followed when the same number is written with numerical signs. This fact has misled some scholars to think that the decimal notation and the word numerals were evolved at two different places. G. R. Kaye has gone so far as to suggest that the word numerals were imported into India from the east. This suggestion is incorrect for the simple reason that in no language other than Sanskrit do we find any early use of the word system. Moreover, in no country other than India do we find any trace of the use of a word system of numeration

[1] *IA*, XXI, p. 48.

[2] The Kadab plates, *IA*, XII, p. 11; declared by Fleet to be suspicious (Kanarese Dynasties; *Bombay Gazatteer*, I, ii, 399, note 7); cf. Bühler, *l.c.*, p. 86, note 4.

[3] The Dholpur Inscription, *Zeitschrift der Deutschen Morgenländischen Gesellschaft*, XL, p. 42.

[4] *IA*, VII, p. 18.

[5] Bühler, *l.c.*, p. 86, note 7.

as far back as the fourth century A.D., at which period it was in common use amongst the astronomers and mathematicians of India.

During the earlier stages of the development of this system, we find that instead of the word symbols, the number names were used, being arranged from left to right just as the numerical signs. An instance of this is found in the Bakhshâlî Manuscript[1] (*c.* 200), where the number

2 6 5 3 2 9 6 2 2 6 4 4 7 0 6 4 9 9 4 . . . 8 3 2 1 8

is expressed as

Ṣaḍviṁśaśca (26) *tripañcâśa* (53) *ekonatriṁśa* (29) *evacha*
Dvâṣa[*ṣṭi*](62) *ṣaḍviṁśa* (26) *catuḥcatvâriṁśa* (44) *saptati* (70)
Catuḥsaṣṭi (64) *na*[*vanavati*] (99) ... *ṁsanantaram*
Trirasiti (83) *ekaviṁśa* (21) *aṣṭa* (8) ... *pakaṃ*

In the same manuscript, however, the contrary arrangement is used when the number 54 is expressed as *catuḥ* (4) *pañca* (5).[2] Jinabhadra Gaṇi (575) has used word symbols with the left to right arrangement to express numbers.[3] It seems, therefore, that in the beginning opinion was divided as to which method of arrangement should be followed in the word system.

The extensive use of the word numerals by early mathematicians such as Puliśa, Varâhamihira, Lalla and others appears to have set the fashion to write the word numerals with a right to left arrangement, which was generally followed by later writers.

[1] Folio 58, recto. The dots indicate some missing figures. The problem apparently required the expression of a big number in numerical denominations. We do not find a problem of this type in any of the later works. Cf. B. Datta, "The Bakhshālī Mathematics." *BCMS*, XXI, p. 21.

[2] Folio 27, recto.

[3] *Bṛhat-kṣetra-samâsa*, i. 69ff.

No explanation as to why the right to left arrangement was preferred in the word system is to be found in any of the ancient works. The following explanation suggests itself to us, and we believe that it is not far from the truth: The different words forming a number chronogram were to be so selected that the resulting word expression would fit in with the metre used. To facilitate the selection the number was first written down in numerical figures. The selection of the proper words would then, naturally, begin with the figure in the units place, and proceed to the left just as in arithmetical operations. This is in accordance with the rule "*aṅkânâṃ vâmato gatiḥ*," *i.e.*, 'the numerals proceed to the left,' which seems to have been very popular with the Indian mathematicians. The right to left arrangement is thus due to the desire of the mathematicians to look upon the process of formation of the word chronogram as a sort of arithmetical operation.

Date of Invention. The use of the word numerals in the *Agni-Purâṇa* which was composed in the 4th century A.D. or earlier, shows that the word system of numerals must have become quite common in India at that time, the *Purâṇas* being works meant for the common folk. That it was a well developed system in the fourth century is also shown by its extensive use in the *Sûrya-siddhânta* and the *Puliśa-siddhânta*. Its invention consequently must be placed at least two centuries earlier. This would give us the period, 100 A.D. to 200 A.D., as the time of its invention. This conclusion is supported by the epigraphic use of the word notation in 605 A.D., in Cambodia, which shows that by the end of the 6th century A.D., the knowledge of the system had spread over an area roughly of the size of Europe.

It must be pointed out here that the decimal place-value notation and the word numerals were not invented at the same time. The decimal notation must have been in existence and in common use amongst the mathematicians long before the idea of applying the place-value principle to a system of word names could have been conceived. Thus we find that in the beginning (*c.* 200), the place-value principle, as is to be expected, was used with the number names. The word symbols were then substituted for the number names for the sake of metrical convenience. The right to left procedure was finally adopted because of the mathematicians' desire to look upon the formation of the word numeral as a sort of mathematical operation.

The above considerations place the invention of the decimal place-value notation at a period, at least two or three centuries before the invention of the word system. The word notation, therefore, points to the 1st century B.C. as the time of invention of the place-value notation. This conclusion agrees with that arrived on epigraphic evidence alone.

11. ALPHABETIC NOTATIONS

The idea of using the letters of the alphabet to denote numbers can be traced back to Pâṇini (*c.* 700 B.C.) who has used the vowels of the Sanskrit alphabet to denote numbers.[1] No definite evidence of the extensive use of an alphabetic notation is, however, found

[1] In Pâṇini's grammar there are a number of sûtras (rules) which apply to a certain number of sûtras that follow and not to all. Such sûtras are marked by signs according to Pâṇini. Patañjali commenting on sûtra i. 3. 11, says that according to Kâtyâyana (4th century B.C.) a letter (*varṇa*), denoting the number of sûtras upto which a particular rule is to apply, is written over the sûtra. Kaiyyaṭa illustrates this remark by saying that the letter *i* is written above Pâṇini's sûtra, v. 1. 30 to show that it applies only to the next two sûtras. Thus according to Pâṇini *a* = 1, *i* = 2, *u* = 3,.....

upto the 5th century A.D. About this period a number of alphabetic notations were invented by different writers with the sole purpose of being used in verse to denote numbers. The word numerals gave big number chronograms, so that sometimes a whole verse or even more would be devoted to the word chronogram only. This feature of the word system was naturally looked upon with disfavour by some of the Indian astronomers who considered brevity and conciseness to be the main attributes of a scientific composition. Thus the alphabetic notations were invented to replace the word system in astronomical treatises. The various alphabetic systems[1] are simple variations of the decimal place-value notation, using letters of the alphabet in the place of numerical figures. It must be noted here that the Hindu alphabetic systems, unlike those employed by the Greeks or the Arabs, were never used by the common people, or for the purpose of making calculations; their knowledge was strictly confined to the learned and their use to the expression of numbers in verse.

Alphabetic System of Aryabhata I. Âryabhaṭa I (499) invented an alphabetic system of notation, which has been used by him in the *Daśagîtikâ*[2] for enumerating the numerical data of his descriptive astronomy. The

[1] Some alphabetic systems used for the pagination of manuscripts do not use the place-value principle. These systems were the invention of scribes who probably wanted to be pedantic and to show off their learning. Their use was confined to copyists of manuscripts.

[2] The *Daśagîtikâ* as the name implies ought to contain ten stanzas, but actually there are thirteen. Of these the first is an invocation to the Gods, the second is the *paribhâṣâ* ("definition") given above and the thirteenth is of the nature of a colophon. These three stanzas are, therefore, not counted. Cf. W. E. Clark, "Hindu-Arabic Numerals," *Indian Studies in Honour of Charles Rockwell Lanman*, (Harvard Univ. Press), 1929, p. 231.

rule is given in the *Daśagītikâ* thus:

Vargâkṣarâṇi varge'varge'vargâkṣarâṇi kât ṅmau yaḥ
Khadvinavake svarâ nava varge'varge navântyavarge vâ

The following translation gives the meaning of the rule as intended by the author:

"The *varga*[1] letters beginning with *k* (are used only) in the *varga*[2] places, the *avarga* letters in the *avarga*[3] places, (thus) *ya* equals *ṅmau* (*ṅa* plus *ma*); the nine vowels (are used to denote) the two nines of zeros of *varga* and *avarga* (places). The same (procedure) may be (repeated) after the end of the nine *varga* places."

This rule has been discussed by Whish,[4] Brockkhaus,[5] Kern,[6] Barth,[7] Rodet,[8] Kaye,[9] Fleet,[10] Datta,[11] Ganguly,[12] Das,[13] Lahiri[14] and Clark.[15]

The translation of *kha* by "place" (Clark) or by "space" (Fleet) is incorrect. We do not find the word *kha* used in the sense of 'notational place' anywhere in Sanskrit literature. Its meanings are 'void', 'sky', etc., and it has been used for zero, in the mathematical and

[1] *Varga* here means "classed," *i.e.*, the classed letters of the alphabet. The first twenty-five letters of the alphabet are classed in groups of five, the remaining ones are unclassed.

[2] *Varga* here means odd.

[3] *Avarga* here means even.

[4] *Transactions of the Literary Society of Madras*, I, 1827, p. 54.

[5] *Zeitschrifte für die kunde des Morgenländes*, IV, p. 81.

[6] *JRAS*, 1863, p. 380.

[7] *Oeuvres*, III, p. 182.

[8] *JA*, 1880, II, p. 440.

[9] *JASB*, 1907, p. 478; *Indian Mathematics*, Calcutta, 1915, p. 30; *The Bakhshālī Manuscript*, Calcutta, 1927, p. 81.

[10] *JARS*, 1911, p. 109.

[11] *Sâhitya-Pariṣad-Patrikâ*, 1929, p. 22.

[12] *BCMS*, 1926, p. 195.

[13] *IHQ*, III, p. 110.

[14] *History of the World* (in Bengali), Vol. IV, p. 178.

[15] *Âryabhaṭîya of Âryabhaṭa*, Chicago, 1930, p. 2.

astronomical works. We thus replace "the two nines of places" in the translation given by Clark by "the two nines of zeros." Clark has given the following reason for not translating *kha* by zero: "That is equivalent to saying that each vowel adds two zeros to the numerical value of the consonant. This, of course, will work from the vowel *i* on; but the vowel *a* does not add two zeros. It adds no zero or one zero depending on whether it is used with *varga* or *avarga* letters. It seems to me, therefore, more likely that a board divided into columns is implied rather than a symbol for zero, as Rodet thinks." The vowels do not add zeros. The explanation will not work for any of the vowels; for instance, *i*, according to this interpretation, would add two zeros to *g* but three zeros to *y*. What really is implied by *kha* is explained by the commentator Sûryadeva as follows: "*khâni śûnyopalakṣitâni, saṅkhyâvinyâsasthânâni teṣâṃ dvinavakaṃ, khadvinavakaṃ, tasmin khadvinavake śûnyopalakṣitdsthânâṣṭâdaśa* (18) *ityarthaḥ*;" that is, "*kha* denotes zero; the places for putting (writing) the numbers are two nines (*dvinavakaṃ*), therefore, *khadvinavake* means the eighteen places denoted by zeros." It may be mentioned here that the Hindus denote the notational places by zeros. Bhâskara I (522), commenting on *Gaṇitapâda*, 2, which gives the names of ten notational places, says:

"*nyâsaśca sthânânâṃ* oooooooooo."

i.e., "writing down the places we have oooooooooo." Bhâskara I is more explicit in the interpretation of *kha* by zero, for in his comments on the above rule, he states: "*khadvinavake svarâ nava varge*: *kha* means zero (*śûnya*). In two nines of zeros (*kha*), so *khadvinavake*; that is, in the eighteen (places) marked by zeros;....."[1]

[1] Commentary on the *Daśagîtikâ* by Bhâskaıa I, "*khadvinavake svarâ nava varge khâni śûnyâni, khânâṃ dvinavakaṃ tasmin khadvinavake aṣṭâdaśa śûnyâkṣiteṣu*......"

Thus *kha* must be translated by zero, although the *kha* (zero) here is equivalent to the 'notational places.'[1] What is implied here is certainly the symbol for the zero and not a board divided into columns.

Clark finds great difficulty in translating *navântya-varge vâ.* The reading *hau* instead of *vâ* suggested by Fleet is not acceptable. The translation given by us accords with the several commentaries (by Bhâskara I, Sûryadeva, Parameśvara and Nîlakaṇṭha) consulted by us. They all agree.

Explanation. Ârayabhaṭa's rule gives the method of expressing the alphabetic chronogram in the decimal place-value notation, and *vice versa.* The notational places are indicated as follows:

au		*o*		*ai*		*e*		*ḷ*		*ṛ*		*u*		*i*		*a*	
a	v	a	v	a	v	a	v	a	v	a	v	a	v	a	v	a	v
o	o	o	o	o	o	o	o	o	o	o	o	o	o	o	o	o	o

where v stands for *varga* and a for *avarga.*

It will be observed that the eighteen places are denoted by zeros and they are divided into nine pairs, each pair consisting of a *varga* place and an *avarga* place, *i.e.*, odd place and even place.[2] The *varga* letters *k* to *m*[3] are used in *varga* places, *i.e.*, odd places only, and denote the numbers 1, 2,........, 25 in succession. The

[1] Nîlakaṇṭha says: "*khadvinavake,* that is, there are eighteen places, the nine *varga* places and the nine *avarga* places ......" See *Âryabhaṭîya*, ed. by K. Sambasiva Sastri, Trivandrum, 1930, p. 6.

[2] The later Indian treatises use the terms *viṣama* and *sama* for *varga* and *avarga* respectively. *Varga* is also used for a square number or the figure.

[3] These are called *varga* or classified letters, because they are classified into groups of five each.

avarga letters *y* to *h* are used in the *avarga* places, *i.e.*, even places only, and denote the numbers 3, 4,......,10 successively. The first *varga* and *avarga* places together constitute the first *varga-avarga* pair, and so on. Nine such *varga-avarga* pairs are denoted by the nine vowels in succession. Thus the first *varga-avarga* pair, *i.e.*, the units and the tens places are denoted by *a*; the second *varga-avarga* pair, *i.e.*, the hundreds and thousands places by *i*; and so on. The vowels thus denote places —zeros according to the Indian usage of denoting the places—and have by themselves no numerical value. When attached to a 'letter-number' a vowel simply denotes the place that the number occupies in the decimal place-value notation. For instance, when the vowel *a* is attached to *y*, it means that the number 3 which *y* denotes is to be put in the first *avarga* place, *i.e.*, the tens place. Thus *ya* is equal to 30. On the other hand when *a* is attached to one of the classed letters, it refers it to the first *varga* place, *i.e.*, the units place. Thus *ṅa* is equal to 5 and *ma* is equal to 25 and *ṅma*[1] is equal to 30. Similarly *yi* denotes that the number 3 is to be put in the thousands place whilst *gi* would mean that the number 3 which *g* represents is to be put in the hundreds place (*g* being a *varga* letter). Thus *yi*=3,000, whilst *gi*=300. It is possible that the zeros already written were rubbed out and the corresponding numerical figures as obtained from a given letter chronogram were substituted in their places. This would automatically give zeros in the vacant places. When this is not done and the numbers are written below the zeros indicating the places, then zeros have

[1] When two consonants are together joined to a vowel, the numbers representing both are referred to the same *varga-avarga* pair. They are added together as in this case, *ṅma* = *ṅa* + *ma* = 5 + 25 = 30.

to be written in the places that remain vacant.[1] The same procedure can be applied to express numbers occupying more than eighteen places, by letting the vowels with *anusvâra* denote the next eighteen places, or by means of any other suitable device.

One advantage of this notation is that it gives very brief chronograms. This advantage is, however, more than counterbalanced by two very serious defects. The first of these is that most of the letter chronograms formed according to this system are very difficult to pronounce. In fact, some of these[2] are so complicated that they cannot be pronounced at all. The second defect is that the system does not allow any great variety in the letter chronograms, as other systems do.

Katapayadi System. In this system the consonants of the Sanskrit alphabet have been used in the place of the numbers 1-9 and zero to express numbers. The conjoint vowels used in the formation of number chronograms, have no numerical significance. It gives brief chronograms, which are generally pleasant sounding words. Skilled writers have been able to coin chronograms which have connected meanings. It is superior to that of Âryabhaṭa I, and also to the word system. Four variants of this system are known to have been used in India. It is probably due to this non-uniformity of notation that the system did not come into general use.

[1] Some examples from the *Âryabhaṭîya* (i. 3):

	ṛ		*u*		*i*		*a*		
	o	o	o	o	o	o	o	o	
khyughṛ		*gh*	*y*	*kh*					
		4	3	2	o	o	o	o	=4320000
cayagiyiṅuśuchlṛ	*l*	*ch*	*ś*	*ṅ*	*y*	*g*	*y*	*c*	
	5	7	7	5	3	3	3	6	=5775336

[2] For instance *niśiṣuṇḷkhṣṛ*, *bhadliknukhṛ*, etc.

First Variant: The first variant of the *Kaṭapayâdi* system is described in the following verse taken from the *Sadratnamâlâ*:

Nañâvacaśca śûnyâni saṁkhyâ kaṭapayâdayaḥ
Miśre tûpânta hal saṁkhyâ na ca cintyo halasvaraḥ

"*n*, *ñ* and the vowels denote zeros; (the letters in succession) beginning with *k*, *ṭ*, *p*, and *y*, denote the digits; in a conjoint consonant only the last one denotes a number; and a consonant not joined to a vowel should be disregarded." According to this system, therefore,

1 is denoted by the letters		*k*, *ṭ*, *p*, *y*.
2 ,,	,, ,,	*kh*, *ṭh*, *ph*, *r*.
3 ,,	,, ,,	*g*, *ḍ*, *b*, *l*.
4 ,,	,, ,,	*gh*, *ḍh*, *bh*, *v*.
5 ,,	,, ,,	*ṅ*, *ṇ*, *m*, *ś*.
6 ,,	,, ,,	*c*, *t*, *ṣ*.
7 ,,	,, ,,	*ch*, *th*, *s*.
8 ,,	,, ,,	*j*, *d*, *h*.
9 ,,	,, ,,	*jh*, *dh*.
0 ,,	,, ,,	*ñ*, *n* and vowels standing by themselves.

The consonants with vowels are used in the places of of the numerical figures just as in the place-value notation. Of conjoint consonants only the last one has numerical significance. A right to left arrangement is employed in the formation of chronograms, just as in the word system, *i.e.*, the letter denoting the units figure is written first, then follows the letter denoting the tens figure and so on. The following examples taken from inscriptions, grant plates and manuscripts will illustrate the system:

(1)[1] *râ* (2) - *gha* (4) - *vâ* (4) - *ya* (1) = 1442,

[1] *EI*, VI, p. 121.

4 4 6
(2)[1] *bha - va - ti* = 644,

5 1 3 1
(3)[2] *śa - ktyâ - lo - ke* = 1315,

6 4 3 1
(4)[3] *ta - tvâ - lo - ke* = 1346,

2 3 1 5 6 5 1
(5)[4] *kha - go - ntyâ - nme - ṣa - mâ - pe* = 1565132.

The origin of this system can be traced back to the fifth century A.D. From a remark[5] made by Sûryadeva in his commentary on the *Âryabhaṭîya*, it appears that the system was known to Âryabhaṭa I (499). Its first occurrence known to us is found in the *Laghu-Bhâskarîya* of Bhâskara I (522).[6]

Second Variant: Âryabhaṭa II (950) has used a modification of the above system. In this variant, the consonants have the same values as above, but the vowels whether standing by themselves or in conjunction with consonants have no numerical significance. Also unlike the first variant, each component of a conjoint consonant has numerical value according to its

[1] *IA*, II, p. 360.

[2] *EI*, III, p. 229.

[3] *EI*, III, p. 38.

[4] The date of the commentary of Ṣaḍguruśiṣya on *Sarvânukramaṇî* is given by this chronogram in the *Kaliyuga* Era: It corresponds to 1184 A.D.

[5] Comments on the *paribhâṣâ-sûtra* of the *Daśagîtikâ*. The author remarks:

"*Vargâkṣarâṇâṁ saṁkhyâ pratipâdane, kaṭapayâditvaṃ nañayośca śûnyâpi siddhaṁ tannirâsârthaṁ kât grahaṇaṃ.*"

That is, "the letters *kât* have been used to distinguish it (the method of Âryabhaṭa I) from the *Kaṭapayâdi* system of denoting numbers by the help of the *varga* letters, where *n* and *ñ* are equal to zero."

[6] *LBh*, i. 18.

place. The letters are arranged in the left to right order just as in writing numerical figures.[1] The difference between the two variants may be illustrated by the chronogram *ḍha-ja-he-ku-na-he-t-sa-bhâ.*[2] According to Âryabhaṭa II it denotes 488108674, whereas according to the first variant it would denote 47801884.

Third Variant: A third variant of this system is found in some Pâlî manuscripts from Burma.[3] This is in all respects the same as the first variant except that *s*=5, *h*=6 and *l'*=7. The modification in the values of these letters are due to the fact that the Pâlî alphabet does not contain the Sanskrit *ś* and *ṣ*.

Fourth Variant: A fourth variant of the system was in use in South India, and is known as the Kerala System. This is the same as the first variant with the difference that the left-to-right arrangement of letters, just as in writing numerical figures, is employed.

Aksarapalli. Various peculiarities are found in the forms as well as the arrangement of the numerical symbols used in the pagination of old manuscripts. These symbols are known as the *akṣarapallî*, *i.e.*, the letter system.[4] In this system the letters or syllables of the script in which the manuscript is written are used to denote the numbers. The following list gives the phonetic values of the various numerals as found in old manuscripts:[5]

[1] The notation is explained in *MSi*, i. 2:

Rûpât kaṭapayapûrvâ varṇâ varṇakramâdbhavantyaṅkâḥ
Nñau śûnyaṁ prathamârthe â chede e tṛtîyârthe.

[2] *MSi*, i. 10.

[3] L. D. Barnet, *JRAS*, 1907, pp. 127 ff.

[4] For forms see Tables.

[5] See *PLM*, pp. 107f.

1 = e, sva, rûṁ.
2 = dvi, sti, na.
3 = tri, śrî, maḥ.
4 = ṅka, rṅka, ṅkâ, ṇka, rṇka, ṣka, rṣka,

 (pke), 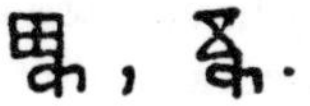 rphra, pu.

5 = tṛ, rtṛ, rtṛâ, hṛ, nṛ, rnṛ.
6 = phra, rphra, rphru, ghna, bhra, rpu, vyâ, phla.
7 = gra, grâ, rgrâ, rgbhrâ, rggâ, bhra.
8 = hra, rhra, rhrâ, dra.
9 = oṁ, ruṁ, ru, uṁ, ûṁ, a, rnuṁ.
10 = ḷ, ḷa, ṇṭa, ḍa, a, rpta.
20 = tha, thâ, rtha, gha, rgha, pva, va.
30 = la, lâ, rla, rlâ.
40 = pta, rpta, ptâ, rptâ, pna.

50 = s, 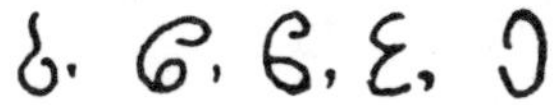 e, i, ṇu.

60 = cu, vu, ghu, thu, rthu, rthû, thû, rgha, rghu.
70 = cû, cu, thû, rthü, rghû, rmta.

80 =  pu.
90 = 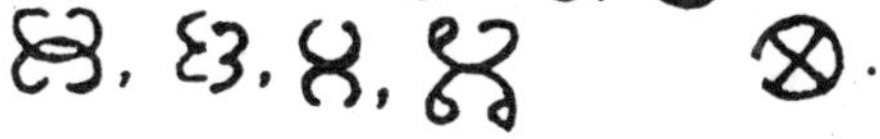

100 = su, sû, lu, a.
200 = sû, â, lû, rghû.
300 = stâ, sûâ, ñûâ, sâ, su, suṁ, sû.
400 = sûo, sto, stâ.

It will be observed that to the same numeral there correspond various phonetical values. Very frequently the difference is slight and has been intentionally made, probably to distinguish the signs with numerical values from those with letter values. In some other cases there are very considerable variations, which (according to Bühler) have been caused by misreadings of older signs or dialectic differences in pronunciation. The symbols are written on the margin of each leaf. Due to lack of space they are generally arranged one below the other in the Chinese fashion. This is so in the Bower manuscript which belongs to the sixth century A.D. In later manuscripts the pages are numbered both in the *akṣarapallî* as well as in decimal figures. Sometimes these notations are mixed up as in the following:[1]

		lâ		su			su
33	=	3 ;	100 =	o ;	102	=	o ;
				o			2
		su		su			sû
131	=	lâ ;	150=	6	209	=	o
		1		o			ruṁ

The *akṣarapallî* has been used in Jaina manuscripts upto the sixteenth century. After this period, the decimal figures are generally used. In Malabar, a system resembling the *akṣarapallî* is in use upto the present day.[2]

[1] Cf. *PLM*, p. 108.

[2] 1 = na, 2 = nna, 3 = nya, 4 = ṣkra,
5 = jhra, 6 = hâ(ha), 7 = gra, 8 = pra,
9 = dre(?), 10 = ma, 20 = tha,
30 = la, 40 = pta, 50 = ba, 60 = tra,
70 = ru (tru), 80 = ca, 90 = ṇa, 100 = ña.
(*Cf. JRAS*, 1896, p. 790)

Other Letter Systems. (*A*) A system of notation in which are employed the sixteen vowels and thirty-four consonants of the Sanskrit alphabet is found in certain manuscripts from Southern India (Malabar and Andhra), Ceylon, Burma and Siam. The thirty-four consonants in order with the vowel *a* denote the numbers from one to thirty-four, then the same consonants with the vowel *â* denote the numbers thirty-five to sixty-eight and so on.[1]

(*B*) Another notation in which the sixteen vowels with the consonant *k* denote the numbers one to sixteen and with *kh* they denote the numbers seventeen to thirty-two, and so on, is found in certain Pâlî manuscripts from Ceylon.[2]

(*C*) In a Pâlî manuscript in the Vienna Imperial Library a similar notation is found with twelve vowels and thirty-four consonants. In this the twelve vowels[3] with *k* denote the numbers from one to twelve, with *kh* they denote the numbers from thirteen to twenty-four, and so on.

These letter systems do not appear to have been in use in Northern India, at least after the third century A.D. They are probably the invention of scribes who copied manuscripts.

12. THE ZERO SYMBOL

Earliest Use. The zero symbol was used in metrics by Piṅgala (before 200 B.C.) in his *Chandaḥ-sûtra*. He gives the solution of the problem of finding the total number of arrangements of two things in *n* places, repetitions being allowed. The two things considered are

[1] Burnell, *South Indian Palaeography*, London, 1878, p. 79.
[2] *Ibid.*
[3] The vowels *ṛ*, *r̤*, *ḷ*, *l̤*, are omitted.

the two kinds of syllables "long" and "short", denoted by *l* and *g* respectively. To find the number of arrangements of long and short syllables in a metre containing *n* syllables, Piṅgala gives the rule in short aphorisms :

"(Place) two when halved;"[1] "when unity is subtracted then (place) zero;"[2] "multiply by two when zero;"[3] "square when halved."[4]

The meaning of the above aphorisms will be clear from the calculations given below for the *Gâyatrî* metre which contains 6 syllables.[5]

	A		B
Place the number	6		
Halve it, result	3	Separately place	2
3 cannot be halved, therefore, subtract 1, result	2	" "	0
Halve it, result	1	" "	2
1 cannot be halved, therefore, subtract 1, result	0	" "	0

The process ends.

The calculation begins from the last number in column B. Taking unity double it at 0, this gives 2; at 2 square this (2), the result is 2^2; then at zero double (2^2), the result is 2^3; ultimately, at 2 square this (2^3), the result is 2^6, which gives the total number of ways

[1] *Piṅgala Chandaḥ-sûtra*, ed. by Sri Sitanath, Calcutta, 1840, viii. 28.

[2] *Ibid*, viii. 29.

[3] *Ibid*, viii. 30.

[4] *Ibid*, viii. 31.

[5] For 7 syllables, the steps are :

Subtract	1	6	place	0	Double	$2.2^6 = 2^7$
Halve		3	"	2	Square	2^6
Subtract	1	2	"	0	Double	$2.2^2 = 2^3$
Halve		1	"	2	Square	2^2
Subtract	1	0	"	0	Double	$1 = 2$

giving 2^7 as the result.

in which two things can be arranged in 6 places.[1]

It will be observed that two symbols are required in the above calculation to distinguish between two kinds of operations, viz., (1) that of halving and (2) that of the 'absence' of halving and subtraction of unity. These might have been denoted by any two marks arbitrarily chosen.[2] The question arises: why did Pingala select the symbols "two" and "zero"? The use of the symbol two can be easily explained as having been suggested by the process of halving—division by the number two. The zero symbol was used probably because of its being associated, at the time, with the notion of 'absence' or 'subtraction.' The use of zero in either sense is found to have been common in Hindu mathematics from early times. The above reference to Piṅgala, however, shows that the Hindus possessed a symbol for zero (*śûnya*), whatever it might have been, before 200 B.C.

The Bakhshâlî Manuscript (*c.* 200) contains the use of zero in calculation. For instance, on folio 56 verso, we have:

"	880	964	multiplied become	848320
	84	168		14112

The square of *forty* different places is | 1600 |. On subtracting this from the number above (numerator), the remainder is | 846720 / 14112 |. On removal of the common factor, it becomes | 60 |."

[1] This method of calculation is not peculiar to the *Piṅgala Chandaḥ-sûtra*. It is found in various other works on metrics as well as mathematics. The zero symbol has been similarly employed in this connection in later works also. *Vide infra.*

[2] *E.g.*, Pṛthudakasvâmî uses *va* (from *varga*, "square") and *gu* (from *guṇa*, "multiply"), while Mahâvîra uses the numerals 1 and 0. *Vide infra.*

There are a large number of passages of this kind in the work. It will be noticed that in such passages the sentences would be incomplete without the figures, so the figures must have been put there at the time of the original composition of the text, and cannot be suspected of being later interpolations. For an explicit reference to zero and an operation with it, we take the following instance from the work:

"

0	2	3	4	visible 200
1	1	1	1	1

[1] Adding[2] unity to zero

1	2	3	......

"[3]

In the *Pañca-siddhântikâ* (505) zero is mentioned at several places. The following is an instance:

"In Aries the minutes are seven, in the last sign six; in Taurus six (repeated) thrice; five (repeated) twice; four; four; in Gemini they are three, two, one, zero (*śûnya*) (each repeated) twice."[4]

Zero is here conceived as a number of the same type as three, two or one. It cannot be correctly interpreted otherwise. Addition and subtraction of zero are also used in expressing numbers in this work for the sake of metrical convenience. For instance:

"Thirty-six *increased* by two, three, nine, twelve, nine, three, zero (*śunya*) are the days."[5]

Instances of the above type all occur in those

[1] The zeros given here are represented in the manuscript by dots. The statement in modern symbols is equivalent to the equation,

$$x + 2x + 3x + 4x = 200.$$

[2] The Sanskrit word is *yutaṁ* meaning literally "adding", but what is meant is "putting" unity for the unknown (zero).

[3] *BMs*, folio 22, verso.

[4] *PSi*, vi. 12.

[5] *PSi*, xviii. 35; other instances of this nature are in iii. 17; iv. 7; iv. 8; iv. 11; xviii. 44; xviii, 45; xviii. 48; xviii. 51.

sections of the *Pañca-siddhântikâ* which deal with the teachings of Puliśa. It seems, therefore, that such expressions are quotations from the *Puliśa-siddhânta*. As it is known that the word numerals were employed by Puliśa (*c*. 400), it can be safely concluded that he was conversant with the concept of the zero as a numeral.

The writings of Jinabhadra Gaṇi (529-589), a contemporary of Varâhamihira, offer conclusive evidence of the use of zero as a distinct numerical symbol. While mentioning large numbers containing several zeros, he often enumerates, obviously for the sake of abridgement, the number of zeros contained. For instance: 224,400,000,000 is mentioned as "twenty-two forty-four, eight zeros;"[1] 3,200,400,000,000 as "thirty-two two zeros four eight zeros."[2] At another place in his work

$$241960\frac{407150}{483920} = 241960\frac{40715}{48392}$$

is described thus :

"Two hundred thousand forty-one thousand nine hundred and sixty; removing (*apavartana*) the zeros, the numerator is four-zero-seven-one-five, and the denominator four-eight-three-nine-two."[3]

It should be noted that the term *apavartana* means what is known in modern arithmetic as the reduction of a fraction to its lowest terms by removing the common factors from the numerator and the denominator. Hence the zero of Jinabhadra Gaṇi is certainly not a mere concept of nothingness but is a specific numerical symbol used in arithmetical calculation.

[1] *Bṛhat-kṣetra-samâsa*, ed. with the commentary of Malayagiri, Bombay, i. 69.

[2] *Ibid*, i. 71. Other such instances are in i. 90, 97, 102, 108, 113, 119, etc.

[3] *Ibid*, i. 83.

Another contemporary mathematician, Bhâskara I (*c.* 525), refers to the subtraction of zero in his *Mahâbhâskarîya*. In his commentary on the *Âryabhaṭîya* he uses the place-value numerals with zero. As has been pointed out before (p. 66) zero is also used by him to denote the notational places.

Siddhasena Gaṇi who lived in the sixth century, has, in his commentary on the *Tattvârthâdhigama-sûtra* of Umâsvâti, used zero in calculation, as is evidenced by the following two typical instances taken from his work:[1]

"....the remainder is this, 3,534,400,000,000. The square-root of this is extracted; *half of the eight zeros are four zeros*; The root of the ramaining portion is one-eight-eight; hence the resulting root is this, 1,880,000."

"On removing the four zeros, the quotient obtained after that is 100,000."

All known Hindu treatises on arithmetic and algebra contain a section dealing with the fundamental operations with zero, including involution and evolution. Details regarding these operations will be given later on; but it must be pointed out here, that these arithmetical operations with zero, certainly presuppose its existence as a numeral denoted by some specific symbol.[2]

[1] *Tattvârthâdhigama-sûtra* of Umâsvâti, with his own gloss, elucidated by Siddhasena Gaṇi, ed. by H. R. Kapadia, Bombay, 1926, iii. 11 (com.).

[2] Smith and Karpinski (*Hindu-Arabic Numerals*, p. 53) state, "..the Gaṇita-Sâra-Saṅgraha of Mahâvirâchârya (c. 830 A.D.), while it does not use the numerals with place value, has a similar discussion with zero." The first part of the statement is incorrect, because Mahâvîra has always used numerals with place-value. In fact, no trace of numerals without place-value is to be found in the *Gaṇita-sâra-saṁgraha*. J. Tropfke's statement (*Geschichte d. Elementar-Mathematik*, Bd. II, 1926, p. 56) that zero was not regard-

Form of the Symbol. The above quotations from ancient works prove conclusively that the zero has been considered as a number in India from the earliest centuries of the Christian era, and that there existed some symbol to denote this number. What exactly was the form of this symbol is doubtful. The Bakhshâlî Manuscript employs a dot for zero, but as the present copy of the work dates from probably the eighth or the ninth century, it cannot be said whether the form of the symbol was the same when the Bakhshâlî work was written, *i.e*, in the third century A.D. or earlier. Evidence as to the use of the dot for zero is also furnished by the writings of Subandhu, a poet litterateur who flourished about the close of the sixth century. In the *Vâsavadattâ* of Subandhu we meet with the following metaphor:

"And at the time of the rising of the moon with its blackness of night, bowing low, as it were, with folded hands under the guise of closing blue lotuses, immediately the stars shone forth,......... like zero dots (*śûnya-bindu*), because of the nullity of metempsychosis, scattered in the sky as if on the ink-blue skin rug of the Creator who reckoneth the sum total with a bit of the moon for chalk."[1]

The term *bindu* ("dot") has been used for zero in word numerals as well as in later literature,[2] when a small circle was in use to denote zero, the dot having

ed as a number before the seventeenth century A.D., is incorrect. *Cf.* B. Datta, "Early literary evidence of the use of the zero in India," *American Math. Monthly*, XXXVIII, 1931, p. 569.

[1] *Vâsavadattâ* of Subandhu, edited by F. Hall (Calcutta, 1859, p. 182) and translated into English by Louis H. Gray (New York, 1913, pp. 99*f*).

[2] *E.g.*, the Hindi poet Bihârî in one of his couplets remarks: "The dot on her forehead is increasing her beauty ten-fold, just as a dot increases a number ten-fold."

been given up long before. The quotation from Subandhu cannot, therefore, be taken as a definite proof of the use of the dot as a symbol for zero in his time. All that we can infer is that at some period before Subandhu, the dot was in use. We may go further and state that very probably, the earliest symbol for zero was a dot and not a small circle.

The earliest epigraphical record of the use of zero is found in the Ragholi plates[1] of Jaivardhana II of the eighth century. The Gwalior inscriptions of the reign of Bhojadeva[2] also contain zero. The form of the symbol in these inscriptions is the small circle. This is the form that has been in common use from quite early times, probably from before the eighth century.

Other Uses of the Symbol. In the present elementary schools in India, the student is taught the names of the several notational places and is made to denote them by zeros arranged in a line. These zeros are written as

. . . . o o o o o o o o o o

The teacher points out the first zero on the right and says 'units', then he proceeds to the next zero saying 'tens' and so on. The student repeats the names after the teacher. This practice of denoting the notational places by zeros can be traced back to the time of Bhâskara I, who, as already pointed out on page 66, in his commentary on the *Âryabhaṭîya*, *Gaṇita-pâda*,2, says:

'Writing down the places, we have

o o o o o o o o o o o."

In all works on arithmetic (*pâṭîgaṇita*) zero has

[1] No. 4 in the list of inscriptions given before.
[2] Nos. 19 and 20 in the list.

been used to denote the unknown. This use of zero can be traced back to the third century A.D. It is used for the unknown in the Bakhshâlî arithmetic. In algebra, however, letters or syllables have been always used for the unknown. It seems that zero for the unknown was employed in arithmetic, really to denote the absence of a quantity, and was not a symbol in the same sense as the algebraic x (*yâ*), for it does not appear in subsequent steps as the algebraic symbols do. This use of zero is mostly found in problems on proportion—the Rule of Five, Rule of Seven, etc. The Arabs also under Hindu influence used zero for the unknown in similar problems. Similar use of zero for the unknown quantity is found in Europe in a Latin manuscript of some lectures by Gottfried Wolack in the University of Erfürt in 1467-68.[1] The dot placed over a number has been used in Hindu *Gaṇita* to denote the negative. In this case it denotes the 'absence' of the positive sign. Similar use of the dot is found in Arabia and Europe obviously under Hindu influence.[2]

13. THE PLACE-VALUE NOTATION IN HINDU LITERATURE

Jaina Canonical Works. The earliest literary evidence of the use of the word "notational place" is furnished by the *Anuyogadvâra-sûtra*,[3] a work written before the Christian era. In this work the total number

[1] Smith and Karpinski, *l.c.*, pp. 53-54.

[2] The occasional use by Al-Battani (929) of the Arabic negative *lâ*, to indicate the absence of minutes (or seconds), noted by Nallino (*Verhandlungen des 5 congresses der Orientalisten*, Berlin, 1882, Vol. II, p. 271), is similar to the use of the zero dot to denote the negative.

[3] The passage has been already quoted in detail (*vide supra* p. 12).

of human beings in the world is given by "a number which when expressed in terms of the denominations, *koṭi-koṭi*, etc., occupies twenty-nine places (*sthâna*)." Reference to the "places of numeration" is found also in a contemporary work, the *Vyavahâra-sûtra*.[1]

Puranas. The Purâṇas which are semi-religious and semi-historical works, also contain references to the notational places. These works were written for the purpose of spreading education on religious and historical matters amongst the common people. Reference to the place-value notation in these works shows the desire of their authors to give prominence to the system. The *Agni-Purâṇa*[2] says:

"In case of multiples from the units place, the value of each place (*sthâna*) is ten times the value of the preceding place."

The *Viṣṇu-Purâṇa*[3] has similarly:

"O *dvija*, from one place to the next in succession, the places are multiples of ten. The eighteenth one of these (places) is called *parârdha*."

The *Vâyu-Purâṇa*[4] observes:

"These are the eighteen places (*sthâna*) of calculation; the sages say that in this way the number of places can be hundreds."

The above three works are the oldest among the Purâṇas and of these the *Agni* and the *Vâyu* Purâṇas in their present form are certainly as old as the fourth century A.D. The *Agni-Purâṇa* is referred by some scholars to the first or second century A.D.

[1] Ch. i.; *cf.* B. Datta, *Scientia*, July, 1931, p. 8.

[2] The *Agni-Purâṇa* contains also the use of the word numerals with place-value (*vide supra* p. 58).

[3] vi. 3.

[4] ci. 102f.

Works on Philosophy. The following simile has been used in *Vyâsa-Bhâṣya*[1] on the *Yoga-sûtra* of Patañjali:

"Thus the same stroke is termed one in the units place, ten in the tens place, and hundred in the hundreds place........"[2]

The same simile occurs in the *Śârîraka-Bhâṣya* of Śaṅkarâcârya:

"Just as, although the stroke is the same, yet by a change of place it acquires the values, one, ten, hundred, thousand, etc...."[3]

The first of the above works cannot be placed later than the sixth century whilst the second one not later than the eighth. The quotations prove conclusively that in the sixth century, the place-value notation was so well known that it could be used as an illustration for a philosophical argument.

Literary Works. A passage from the *Vâsavadattâ* of Subandhu comparing the stars with zero dots has already been mentioned. Several other instances of the use of zero are found in later literature, but they need not be mentioned here.[4]

[1] iii. 13.

[2] The translation is as given by J. H. Woods, *The Yoga System of Patañjali*, p. 216. In a foot-note, it is remarked: "Contrary to Mr. G. R. Kaye's opinion, the following passages show that the place-value system of decimals was known as early as the sixth century A.D." The above passage is also noted by Sir P. C. Ray in his *History of Hindu Chemistry*, Vol. II, p. 117.

[3] III. iii. 17; *cf.* B. Datta, *American Math. Monthly*, XXXIII 1926, pp. 220-1.

[4] *E.g.*, the use of the *śûnya-bindu* in *Naiṣadha-carita* of Śrîharṣa (*c.* 12th century). *Cf.* B. Datta, *Ibid*, pp. 449-454.

14. DATE OF INVENTION OF THE PLACE-VALUE NOTATION

We may now summarise the various evidences regarding the early use of the place-value notation in India:

(1) The earliest palaeographic record of the use of the place-value system belongs to the close of the sixth century A.D.

(2) The earliest use of the place-value principle with the word numerals belongs to the second or the third century A.D. It occurs in the *Agṇi-Purâṇa*, the Bakhshâlî Manuscript and the *Puliśa-siddhânta*.

(3) The earliest use of the place-value principle with the letter numerals is found in the works of Bhâskara I about the beginning of the sixth century A.D.

(4) The earliest use of the place-value system in a mathematical work occurs in the Bakhshâlî Manuscript about 200 A.D. It occurs in the *Âryabhaṭîya* composed in 499 A.D., and in all later works without exception.

(5) References to the place-value system are found in literature from about 100 B.C. Three references ranging from the second to the fourth century A.D. are found in the Purâṇas.

(6) The use of a symbol for zero is found in Piṅgala's *Chandaḥ-sûtra* as early as 200 B.C.

The reader will observe that the literary and non-mathematical works give much earlier instances of the use of the place-value system than the mathematical works. This is exactly what one should expect. The system when invented must have for some time been used only for writing big numbers. A long time must have elapsed before the methods of performing arithmetical operations with them were invented. The system cannot be expected to occur in a mathematical work before it is in a perfect form. Therefore, the evidences furnished by non-mathematical works should, in fact, be earlier than those of mathematical works.

Mathematical works are not as permanent as religious or literary works. The study of a particular mathematical work is given up as soon as another better work comes into the field. In fact, a new mathematical work is composed with a view to removing the defects of and superseding the older ones. It is quite probable that works employing the place-value notation were written before Âryabhaṭa I, but they were given up and are lost. It will be idle to expect to find copies of such works after a lapse of sixteen hundred years.

In Europe and in Arabia it is still possible to find mss. copies of works using the old numerals or a mixture of the old numerals with the new place-value numerals, but in India absolutely no trace of any such work exists.

In Europe the first definite traces of the place-value numerals are found in the tenth and eleventh centuries, but the numerals came into general use in mathematical text books in the seventeenth century. In India Âryabhaṭa I (499), Bhâskara I (522), Lalla (*c.* 598), and Brahmagupta (628), all use the place-value numerals. There is no trace of any other system of notation in their works. Following the analogy of Europe, we may conclude, on the evidence furnished by Hindu mathematical works alone, that the place-value system might have been known in India about 200 B.C.

As the literary evidence also takes us to that period, we may be certain that the place-value system was known in India about 200 B.C. Therefore we shall not be much in error, if we fix 200 B.C. as the probable date of invention of the place-value system and zero in India. It is possible that further evidence may force us to fix an earlier date.

15. HINDU NUMERALS IN ARABIA[1]

The regular history of the Arabs begins after the flight of Mohammad from Mecca to Medina in A.D. 622. The spread of Islam succeeded in bringing together the scattered tribes of the Arabian Peninsula and creating a powerful nation. The united Arabs, within a short space of time, conquered the whole of Northern Africa and the Spanish Peninsula, and extended their dominions in the east upto the western border of India. They easily put aside their former nomadic life, and adopted a higher civilisation.

The foundations of Arabic literature and science were laid between 750-850 A.D. This was done chiefly with the aid of foreigners and with foreign material. The bulk of their narrative literature came to the Arabs in translation from Persian. Books on the science of war, the knowledge of weapons, the veterinary art, falconry, and the various methods of divination, and some books on medicine were translated from Sanskrit and Persian. They got the exact sciences from Greece and India.

Before the time of Mohammad the Arabs did not possess a satisfactory numeral notation. The numerous computations connected with the financial administration of the conquered lands, however, made the use of a developed numeral notation indispensable. In some localities the numerals of the more civilised conquered nations were used for a time. Thus in Syria, the Greek notation was retained, and in Egypt the

[1] For details consult Cajori's *History of Mathematics*, and Smith and Karpinski's *Hindu Arabic Numerals*.

Coptic. To this early period belongs the Edict of Khalif Walid (699) which forbade the use of the Greek language in public accounts, but made a special reservation in favour of Greek letters as numerical signs, on the ground that the Arabic language possessed no numerals of its own.[1] The Arabic letters gradually replaced the Greek ones in the alphabetic notation and the *abjad* notation came to be used. It is probable that the Arabs had come to know of the Hindu numerals from the writings of scholars like Sebokht, and aslo of their old *ghobâr* forms from other sources. But as their informants could not supply all the necessary information (*e.g.*, the methods of performing the ordinary operations of arithmetic) these numerals had to wait for another century before they were adopted in some of their mathematical works.

During the reign of the Khalif Al-Mansûr (753-774 A.D.) there came embassies from Sindh to Baghdad, and among them were scholars, who brought along with them several works on mathematics including the *Brâhma-sphuṭa-siddhânta* and the *Khaṇḍa-khâdyaka* of Brahmagupta. With the help of those scholars, Al-fazârî, perhaps also Yâkub ibn Târik, translated them into Arabic. Both works were largely used and exercised great influence on Arab mathematics. It was on that occasion that the Arabs first became acquainted with a scientific system of astronomy. It is believed by all writers on the subject that it was at that time that the Hindu numerals were first definitely introduced amongst the Arabs. It also seems that the Arabs at first adopted the *ghobâr* forms of the numerals, which they had already obtained (but without zero) from the

[1] Theophanes (758-818 A.D.), "Chronographia;" *Scriptores Historiae Byzantinae*, Vol. XXXIX, Bonnane, 1839, p. 575; quoted by Smith and Karpinski, *l.c.*, p. 64, note.

Alexandrians, or from the Syrians who were employed as translators by the Khalifs at Baghdad. Al-Khowârîzmî (825), one of the earliest writers on arithmetic among the Arabs, has used the *ghobâr* forms.[1] But not long afterwards, the Arabs realised that the *ghobâr* forms were not suited to their right-to-left script. Then there appears to have been made an attempt to use more convenient forms. But as people had got accustomed to the *ghobâr* forms, they did not like to give them up, and so we find a struggle[2] between the two forms, which continued for about two centuries (10th and 11th) until at last the more convenient ones came into general use. The west Arabs on the other hand did not adopt the modified forms of the east Arabs, but continued to use the *ghobâr* forms, and were thus able to transmit them to awakening Europe. This, perhaps, explains in a better way the divergence in the forms of modern Arabic and modern European numerals, than any theory yet propounded.

In a theory that was advanced by Woepcke, this divergence is explained by assuming that (1) about the second century after Christ, before zero had been invented, the Hindu numerals were brought to Alexandria, whence they spread to Rome and also to west Africa; (2) that in the eighth century, after the notation in India had been already much modified and perfected

[1] Smith and Karpinski, *l.c.*, p. 98.

[2] One document cited by Woepcke is of special interest since it shows the use of the ordinary Arabic forms alongside the *ghobâr* at an early date (970 A.D.). The title of the work is "Interesting and Beautiful Problems on Numbers" copied by Ahmed ibn Mohammed ibn Abdaljalîl Abû Sa'id, al-Sijzî, (951-1024) from a work by a priest and physician, Nazîf ibn Yumn, al-Qass (died 990). Sprenger also calls attention to this fact (in *Zeit. d deutschen-morgenländischen Gesselschaft*, XLV, p. 367). Ali ibn Ahmed Al-Nasâvî (*c.* 1025) tells us that the symbolism of numbers was unsettled in his day (Smith and Karpinski, *l.c.*, p. 98).

by the invention of zero, the Arabs at Baghdad got it from the Hindus; (3) that the Arabs of the west borrowed the Columbus-egg, the zero, from those in the east but retained the old forms of the nine numerals, if for no other reason, simply to be contrary to their political enemies of the east; (4) that the old forms were remembered by the west Arabs to be of Hindu origin, and were hence called *ghobâr* numerals; (5) that, since the eighth century, the numerals in India underwent further changes and assumed the greatly modified forms of modern Devanâgarî numerals.

Now, as to the fact that these figures might have been known in Alexandria in the second century A.D., there is not much doubt. But the question naturally arises: Why should the Alexandrians use and retain a knowledge of these numerals? As far as we know, they did possess numeral notations of their own; why should they give preference to a foreign notation? These questions cannot be satisfactorily answered unless we assume that along with the nine symbols the principle of place-value and probably also the zero was communicated to them. But as they were unprepared for the reception of this abstract conception, they adopted the nine numerals only and used them on the apices. These numerals were then transmitted by them to Rome and to west Africa.

The second assumption that the Hindu numeral figures of the eighth century were adopted by the Arabs is not supported by fact. The figures that are found in the old Arabic manuscripts resemble either the *ghobâr* numerals or the modern Arabic more than the Hindu numerals of the eighth century. In fact, we have every reason to believe that the Arabs knew these *ghobâr* forms, perhaps without the principle of place-value and zero, long before they had direct contact with India, and that they adopted zero only about 750 A.D.

16. HINDU NUMERALS IN EUROPE

Boethius Question. It cannot be definitely said when and how the Hindu numerals reached Europe. Their earliest occurrence is found in a manuscript of the Geometry of Boethius (*c.* 500), said to belong to the tenth century. There are several other manuscripts of this work and they all contain the numerals. Some of these contain the zero whilst the others do not. If these manuscripts (or the portions of them that contain the numerals) be regarded as genuine, it will have to be acknowledged that the Hindu numerals had reached Southern Europe about the close of the fifth century. There are some who consider the passages dealing with the Hindu numerals in the Geometry of Boethius to be spurious. Their arguments can be summarised as below :

(1) The passages in question have no connection with the main theme of the work, which is geometry. The Hindu numerals have not been mentioned in the Arithmetic of Boethius. They have not been used by him anywhere else. Neither Boethius' contemporary Capella (*c.* 475), nor any of the numerous mediaeval writers who knew the works of Boethius makes any reference to the numerals.

(2) The Hindu numeral notation was perfected in India much later than the fifth century, so that the numerals, even if they had been taken to Europe along the trade routes, had no claim to any superiority over the numerals of the west, and so could not have attracted the attention of Boethius.

Of the above arguments, the second is against facts, for it is now established that the Hindu numeral notation with zero was perfected and was in use in India during the earliest centuries of the Christian era. The numerals could have, therefore, easily reached

Europe along the trade routes in the fifth century or even earlier. The first argument is purely speculative and throws doubt on the authenticity of the occurrence of the numerals in Boethius' Geometry. It does not prove anything. It seems to us unfair to question the genuineness of the occurrence of the numerals, when they are found in all manuscripts of the work that are in existence now. Their occurrence in the Geometry can be easily explained on the ground that Boethius' knowledge of those numerals was very meagre. He had obtained the forms from some source—from the Neo-Pythagoreans or direct from some merchant or wandering scholar—but did not know their use. He might have known their use in writing big numbers by the help of the principle of place-value and zero, but he certainly did not know how the elementary operations of arithmetic were to be performed with those numerals. Hence he could make no use of them in his arithmetic or any other work. The writings of Sebokht (*c.* 650) show that the fame of the numerals had reached the west long before they were definitely introduced there. The question of the introduction of the Hindu numerals through the agency of Boethius may, therefore, be regarded as an open one, until further investigations decide it one way or the other.

Definite Evidence. The first writer to describe the *ghobâr* numerals in any scientific way in Christian Europe was Gerbert, a French monk. He was a distinguished scholar, held high ecclesiastical positions in Italy, and was elected to the Papal chair (999). He had also been to Spain for three years. It is not definitely known where he found these numerals. Some say that he obtained them from the Moors in Spain, while others assert that he got them from some other source, probably through the merchants. We find that Gerbert did not appreciate these numerals (and rightly, for there was neither zero nor the place-value), and that in his works, known as the *Regula de abaco computi and the Libellus*, he has used the Roman forms. We thus see that upto the time of Gerbert (died 1003) the principle of place-value was not known in Europe.

As early as 711 A.D., the power of the Goths was shattered at the battle of Jarez de le Frontera, and immediately afterwards the Moors became masters of Spain, and remained so for five hundred years. The knowledge of the modern system of notation which was definitely introduced at Baghdad about the middle of the eighth century must have travelled to Spain and from there made its way into Europe. The schools established by the Moors at Cordova, Granada, and Toledo were famous seats of learning throughout the middle ages, and attracted students from all parts of Europe. Thus although Europe may not be directly indebted to the Moors for its numerical symbols, it certainly is for that important principle which made the ordinary *ghobâr* forms superior to the Roman numerals.

Several instances of the modern system of notation are to be found in Europe in the twelfth century, but no definite attempt seems to have been made for popularizing it before the thirteenth century. Perhaps the most influential in spreading these numerals in Europe was Leonardo Fibonacci of Pisa. Leonardo's father was a commercial agent at Bugia, the modern Bougie, on the coast of Barbary. It had one of the best harbours, and at the close of the twelfth century was the centre of African commerce. Here Leonardo went to school to a Moorish master. On attaining manhood he started on a tour of the Mediterranean and visited Egypt, Syria, Greece, Italy and Provence, meeting with scholars and merchants and imbibing a knowledge of the various systems of numbers in use in the centres of trade. All these systems, he however says, he counted as errors compared with that of the Hindus.[1] Returning to Pisa he wrote his *Liber Abaci* in 1202, rewriting it in 1228.[2] In this the Hindu numerals are explained and used in the usual computations of business. At first Leonardo's book met with a cold reception from the public, because it was too advanced for the merchants and too novel for the universities. However, as time went on people began to realise its importance, and then we find it occupying the highest place among the mathematical classics of the period.

Among other writers whose treatises have helped the spread of the numerals may be mentioned Alexander de Villa Die (*c.* 1240) and John of Halifax (*c.* 1250).

A most determined fight against the spread of these numerals was put up by the abacists who did not use zero but employed an abacus and the apices. But the writings of men like Leonardo succeeded in silencing them, although it took two or three centuries to do so. By the middle of the fifteenth century we find that these numerals were generally adopted by all the nations of western Europe, but they came into common use only in the seventeenth century.

17. MISCELLANEOUS REFERENCES TO THE HINDU NUMERALS

Syrian Reference. The following reference to a passage[3] in a work of Severus Sebokht (662) shows that the fame of the Hindu numerals had reached the banks of

[1] "Sed hoc totum et algorismum atque arcus pictagore quasi errorem computavi respectu modi indorum."

[2] Smith and Karpinski, *l. c.*, p. 131.

[3] Attention was first drawn to this passage by F. Nau. *JA*, II, 1910, pp. 225-227; also see J. Ginsburg, *Bull. American Math. Soc.*, XXIII, 1917, p. 368.

the Euphrates in the beginning of the seventh century. Sebokht was a learned Syrian scholar who lived in the convent of Kenneshre on the Euphrates, in the time of the patriarch Athanasius Gamenale (died 631) and his successor John.[1] Sebokht seems to have been hurt by the arrogance of certain Greek scholars who looked down on the Syrians, and in defending the latter, he claims for them the invention of astronomy. He asserts that the Greeks were merely the pupils of the Chaldeans of Babylon and claims that these same Chaldeans were the very Syrians whom his opponents condemned. It is in this connection that he mentions the Hindus by way of illustration, using the following words:

"I will omit all discussion of the science of the Hindus, a people not the same as the Syrians, their subtle discoveries in the science of astronomy, discoveries that are more ingenious than those of the Greeks and the Babylonians; their computing that surpasses description. I wish only to say that this computation is done by means of nine signs. If those who believe because they speak Greek, that they have reached the limits of science should know these things, they would be convinced that there are also others who know something."

This fragment clearly shows that not only did Sebokht know something of the numerals but that he understood their full significance, and may even have known the zero as Rabbi ben Ezra did (*vide infra* p. 103), in spite of the fact that he too speaks of nine numerals. In fact even today, we speak of nine numerals only, zero not being recognised a numeral.

Arabic References. *Ibn Wahshiya* (855). In his work, *Ancient alphabets and hieroglyphic characters explained*, etc., Ibn Wahshiya gives three forms of

[1] W. Wright, *Short History of Syriac Literature*, pp. 137-139.

Hindu numerals as three species of Hindu alphabets.[1] It seems that the forms were well known in his time in Arabia. In the first set of these forms, dots are used as zeros in the now customary way among the Arabs; in the second set the dots are superposed,[2] while in the third loops are attached to the ordinary symbols for the units.[3] Karpinski thinks that the use of loops was probably the invention of Waḥshiya.

Djâhiz (died 869)

The Arab philosopher Djâhiz calls the numerals "figures of Hind" and observes that with these numerals large numbers can be represented with great facility. He asks, "Who is the inventor of the figures of Hind, who are the authors of the *Sindhind* and of the *Arkhand* and the methods of calculation with the ciphers?"[4]

Abul Hasan Al-Masûdî (943)

The Arab historian Al-Masûdî writes, "A congress of sages at the command of the Creator Brahmâ invented the nine figures and also their (the Hindus') astronomy and other sciences."[5] Al-Masûdî's evidence is important as he had first-hand information of India. He was in Multan in 912 A.D. and in Cambay in 916 A.D.

[1] Karpinski, *Bibl. Math.*, XIII, 1912-13, pp. 97-8.

[2] Similar use of dots is found in certain manuscripts in India. Vide R. L. Mitra, *Notices of Sanskrit MSS.*, V, p. 299, MS. No. 1976, Plate No. I; *India Office Catalogue*, MS. Nos. 1946, 1947 and 1871.

[3] This form is similar to the use of ligatures in India when the numerals were written without place-value, (*vide supra*).

[4] Djâhiz's work is available in French translation (Caire, *Du rond et du carré*, 1324). Cf. Carra de Vaux, *Scientia*, XXI, 1921, pp. 273-282.

[5] Reinaud, *Mémoire sur l'Inde*, Paris, 1894, p. 300. This shows that the Hindus in the tenth century had forgotten the inventor or inventors of the numerals, as a very long time had elapsed.

Abû Sahl Ibn Tamim (950)

Ibn Tamim, a native of Kairwân, a village in Tunis in the north of Africa, wrote in his commentary on the *Sefer Yesîrâh*: "The Indians have invented the nine signs for marking the units. I have spoken sufficiently of them in a book that I have composed on the Hindu calculation, known under the name of *Hisâb al-ghobâr*."[1]

Al-Nadîm (987)

In the *Fihrist*, the author Al-Nadîm includes the Hindu numerals in a list of some two hundred alphabets of India (*Hind.*) These numerals are called *hindisah*.[2]

Al-Bîrûnî (1030)

Al-Bîrûnî resided in India for nearly thirteen years (1017-1030) and devoted himself to the study of the arts and sciences of the Hindus. He had also a remarkable knowledge of the Greek sciences and literature, so he was more qualified than any contemporary or even anterior Arab writer to speak with authority about the origin of the numerals. He wrote two books, viz., *Kitâb al-arqam* ("Book of Ciphers") and *Tazkira fi al-hisâb w'al-madd bi al-arqam al-Sind w'al-Hind* ("A treatise on arithmetic and the system of counting with the ciphers of Sindh and India"). In his *Tarikh al-Hind* ("Chronicles of India"), he says: "As in different parts of India the letters have different shapes, the numerical signs too, which are called *aṅka*, differ. The numerical signs which we use are derived from the finest forms of the Hindu signs."[3] At another place he remarks: "The Hindus use the

[1] Reinaud, *l.c.*, p. 399.

[2] *Kitâb al-Fibrist*, ed. G. Flügel, II, pp. 18-19.

[3] *Alberuni's India*, English translation by E. C. Sachau, London 2nd ed., 1910, Vol. I, p. 74.

numeral signs in arithmetic in the same way as we do. I have composed a treatise showing how far possibly, the Hindus are ahead of us in this subject."[1] In his *Athâr-ul-Bâkiya*[2] ("Vestiges of the Past," written in 1000 A.D.) Al-Bîrûnî calls the modern numerals as *al-arqam al-hind*, *i.e.*, "the Indian Ciphers" and he has incidentally referred to their distinction from two other systems of expressing numbers, *viz.*, the sexagesimal system and the alphabetic system (*Harûf al-jumal*).

Abenragel (1048)

It has been stated by Ali bin Abil-Regal Abul-Hasan, called Abenragel, in the preface to his treatise on astronomy that the invention of reckoning with nine ciphers is due to the Hindu philosophers.[3]

Saraf-Eddin (1172)

Mahmûd bin Qajid al-Amûnî Saraf-Eddin of Mecca wrote a treatise, entitled *Fi al-handasa w'al arqam al-hindi* ("On geometry and the Indian ciphers").[4]

Alkalasâdî (died 1486)

In his commentary of the *Talkhis* of Ibn Albanna, Abul Hasan Ali Alkalasâdî states: "These nine signs, called the signs of the *ghobâr* (dust), are those that are employed very frequently in our Spanish provinces and in the countries of Maghrib and of Africa. Their origin is said to have been attributed by tradition to a man of the Indian nation. This man is said to have taken some fine dust, spread it upon a table and taught

[1] *Ibid*, I, p. 177.

[2] *The Chronology of Ancient Nations*, ed. by Sachau, London, 1879, pp. 62 and 132.

[3] J. F. Montucla, *Historie des Mathématiques*, vol. I, p. 376.

[4] H. Suter, *Die Mathematiker und Astronomer der Ararbe und ihre Werke*, Leipzig, 1900, p. 126.

the people multiplication, division and other operations."[1]

Behâ Eddin (*c.* 1600)

Referring to the numerals Behâ Eddin observes: "The Hindu savants have, in fact, invented the nine known characters."[2]

In the quotations from Arab scholars given above, the term *Hind* has been used for India, and *Hindi* for Indian. *Hind* is the term generally used in Arabic and Persian literature for India. In early writings distinction was sometimes made between Sind and Hind. Thus Al-Masûdî and Al-Bîrûnî used Sind to denote the countries to the west of the river Indus. This distinction is clearly in evidence in Ibn Hawkal's map, reproduced in Elliot and Dawson's *History of India*. There were others who did not make this distinction. Thus Istakri (912) uses Hind to denote the whole of India.[3] Again in the *Shâhnâmâ* of Firdausi,[4] Sind has been used for a river as well as for a country, and Hind for the whole of India. In later times this distinction disappeared completely. According to the lexicographers Ibn Seedeh (died 1066) and Firouzâbâdi (1328-1413), Hind is "the name of a well known nation" and according to El-Jowharee (1008) it denotes "the name of a country." Instances of the use of *Hind* to denote India in the literature of the Arabs can be multiplied at pleasure.

Carra de Vaux[5] has suggested that the word *Hind*

[1] *JA*, I, 1863, pp. 59f.

[2] *Kholasât al-hisâb*, translated into French by A. Marre, *Nouv. Ann. Math.*, V, 1864, p. 266.

[3] Elliot and Dawson's *History of India*, II, p. 412.

[4] English translation by A.G. Warner and E. Warner, London, 1906.

[5] Carra de Vaux, *Scientia*, XXI, 1917, p. 273.

does not probably mean India but is really derived from *ènd* (or *hènd*) signifying "measure," "arithmetic" or "geometry." He concludes that the expression "the signs of *hind*" means "the arithmetical signs" and not "the signs of India." As regards the use of the adjective *hindi* by certain scholars in connection with the numerals, he conjectures that it has probably been employed through confusion for *hindasi*.

Carra de Vaux's derivation of the word *hind* from *ènd* or *hènd* cannot be accepted. It has no support from Arabic lexicography. Moreover, the word *hind* is a very ancient one. It occurs in the *Avesta*[1] both in the earlier *Yasna* and in the later (Sassanian) *Vendidad*. The word also occurs in the cuneiform inscriptions of Darius Hystaspes. The Pehlavi writings before the Arab conquest of Iran also show the word *hind*. In all those cases it means India.

The word *hindi* is an adjective formed from *hind* and means "Indian". The fact that in a few isolated cases, it has been confused with the word *hindasi*, cannot make us conclude that this has happened in all cases.

The terms Hindasa, etc. The words *hindasa*, *hindisa*, *handasa*, *hindasi*, *handasi*, etc., have been stated by competent authorities to be adjectives formed from *hind*, meaning "Indian". Kaye[2] and Carra de Vaux[3] oppose this interpretation. Relying on the lexicon of Firouzâbâdi they assert that these terms are derived from the Persian *andâzah*, meaning "measure." There is no doubt that the word *hindasi* denotes "geometrical" in the Arabic language. But

[1] *Yasna*, x. 141; *Yt.*, x. 104 (*Mihir Yast*).

[2] Kaye, *JASB*, III, 1907, p. 489, also *JASB*, VII, 1911, pp. 810f.

[3] Carra de Vaux, *l.c.*

when this term is used in connection with an explanation of the rule of "double false position" or the method of "proof by nines" or in connection with the "numeral notation," we have to admit that it had some other significance also. As the arithmetical rules designated by the term *hindasi* are found in Hindu arithmetic prior to their occurrence in Arabia, it follows that *hindasi* also means Indian. The term *hindasi*, *hindasa* or *handasa* has, therefore, two meanings, one "geometrical" and the other "Indian". The controversy regarding the meaning of this term which was set at rest by Woepcke,[1] has arisen again because Kaye and Carra de Vaux have refused to recognise both meanings of this term.[2] It may be pointed out here that as one of the meanings of *hindasi* is synonymous with *hindi*, there is no wonder that the two words were sometimes confused with each other, especially by scribes who did not understand the text.

European References. *Isidorus of Seville.* The nine characters (of the *ghobâr* type), without zero, are given as an addition to the first chapter of the third book of the *Origines* by Isidorus of Seville in which the Roman numerals are under discussion. Another Spanish copy of the same work (of 992 A.D.) contains the numerals in the corresponding section. The writer ascribes an Indian origin to them in the following words: "Item de figuris arithmetice. Scire debemus in Indos subtilissimum ingenium habere et ceteris gentes eis in arithmetica et geometria et ceteris liberalibus disciplinis concedere. Et hoc manifestum

[1] Woepcke, *JA*, I, 1863, pp. 27f. See also Suter's article on *handasa* in the *Encyclopaedia of Islam* and Rosen's *Algebra of Mohammad Ben Musa*, London, 1831, pp. 196f.

[2] It will not be difficult to point out in any literature words having more than one meaning. Occasionally these meanings have no connection. Whenever such a word is used, the appropriate meaning has to be deduced from the context.

est in nobem figuris, quibus designant unum-quemque gradum cuinslibet gradus. Quarum hec sunt forma."[1]

Rabbi ben Ezra (1092-1167)

Rabbi Abraham ibn Meir ibn Ezra in his work, *Sefer ha-mispar* ("the Book of Number"), gives the Hindu forms of the numerals. He knew of the Hindu origin of the numerals for he states: "that is why the wise people of India have designated all their numbers through nine and have built forms for the nine ciphers."[2]

Leonardo of Pisa

Leonardo of Pisa in his work, *Liber Abaci* (1202), frequently refers to the nine Indian figures. At one place he says: "Ubi ex mirabili magisterio in arte per novem figuras indorum introductus" etc. In another place, as a heading to a separate division, he writes "De cognitione novem figurarum yndorum" etc., "Novem figure indorum he sunt 9 8 7 6 5 4 3 2 1."[3]

Alexander de Villa Dei

Alexander de Villa Dei (*c.* 1240) wrote a commentary on a set of verses called *Carmen de Algorismo.* In this commentary he writes: "This boke is called the boke of algorim or augrym after lewder use. And this boke tretys of the Craft of Nombryng, the quych crafte is called also algorym. Ther was a kyng of Inde the quich heyth Algor & he made this craft...... Algorisms, in the quych we use teen figurys of Inde."[4]

[1] Quoted by Smith and Karpinski, *l.c.*, p. 138.

[2] *Sefer ha-Mispar, Das Buch der zahl, ein hebraisch-arithmetisches Work des R. Abraham ibn Esra*, Moritz Silberberg, Frankfurt a M., 1895, p. 2.

[3] *Liber Abaci*, Rome, 1857; quoted by Smith and Karpinski, *l.c.*, p. 10.

[4] Smith and Karpinski, *l.c.*, p. 11.

Maximus Planudes (*c.* 1330)

Maximus Planudes states that "the nine symbols come from the Indians."[1]

[1] Wäschke's German Translation, Halle, 1878, p. 3.

TABLE I—*Kharoṣṭhî Numerals*

Śaka, Pârthian and Kuṣâna Inscriptions			Aśoka Inscs.	
𐩅𐩅	40	𐩀	𐩀	1
𐩄𐩅𐩅	50	𐩁	𐩁	2
𐩅𐩅𐩅	60	𐩂		3
𐩄𐩅𐩅𐩅	70	𐩃	////	4
𐩅𐩅𐩅𐩅	80	𐩀𐩃	/////	5
𐩆𐩀	100	𐩁𐩃		6
𐩆𐩁	200	𐩂𐩃		7
𐩆𐩂	300	𐩃𐩃		8
𐩁𐩅𐩆𐩀	122	𐩄		10
𐩃𐩄𐩅𐩅𐩅𐩆𐩁	274	𐩅		20

TABLE II(*a*)—*Semitic and Brâhmî Numerals*

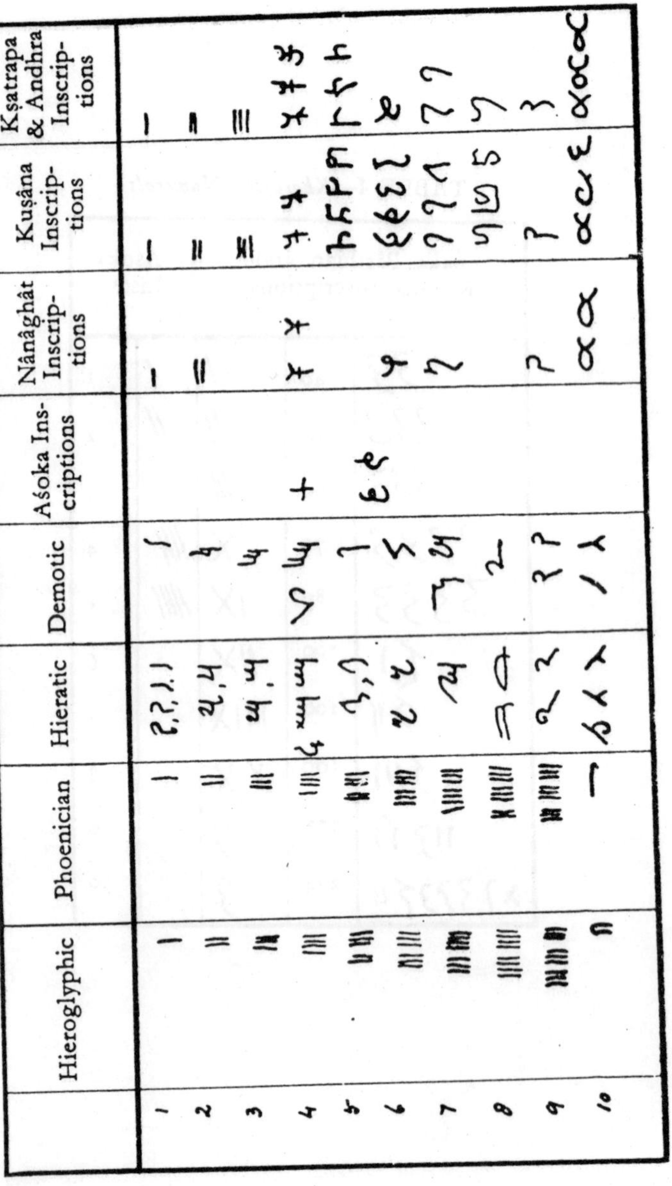

TABLE II(*b*)—*Semitic and Brāhmī Numerals*

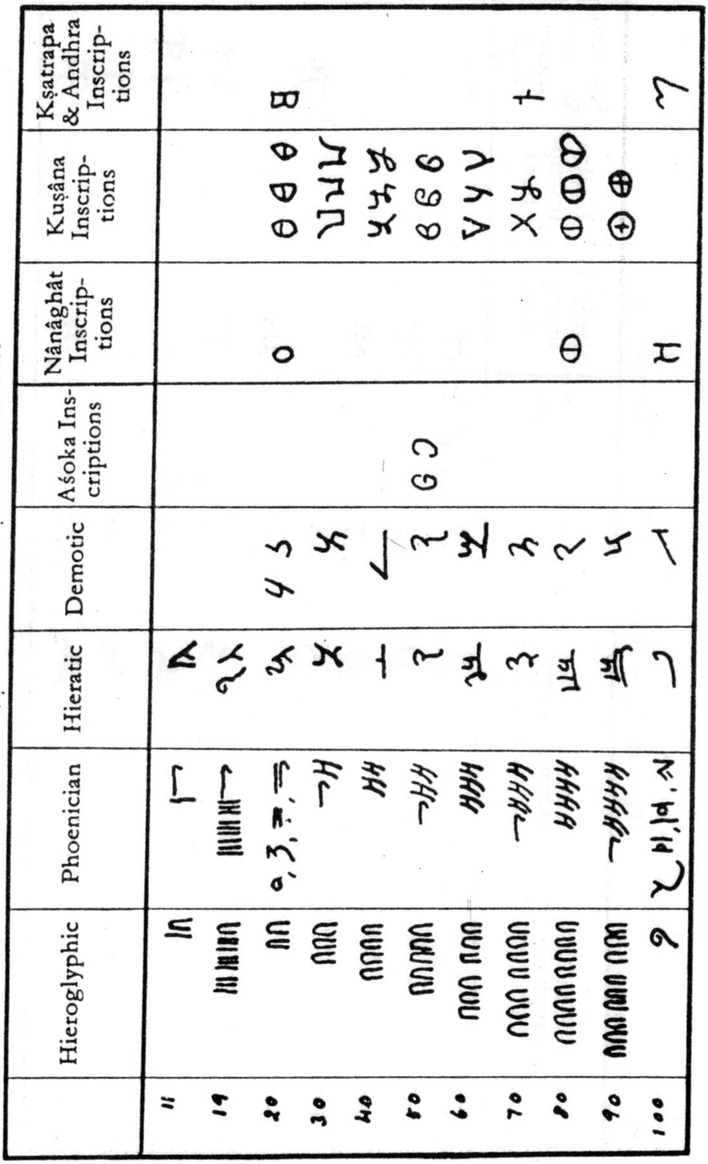

TABLE II(*c*)—*Semitic and Brâhmî Numerals*

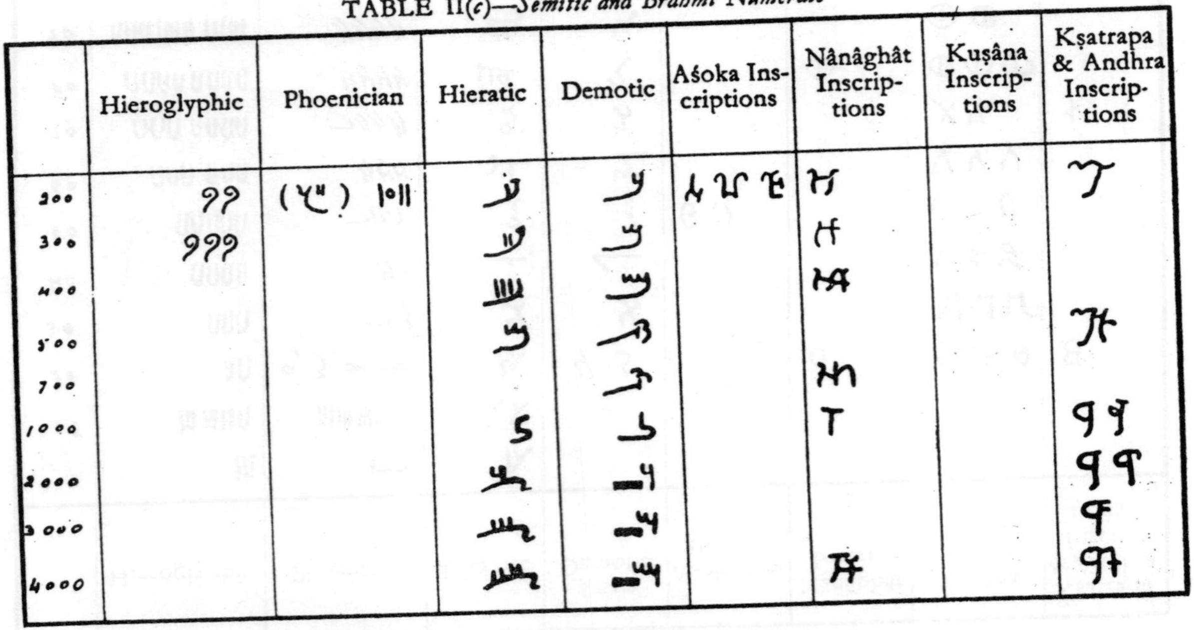

TABLE III—*Brâhmî Numerals*

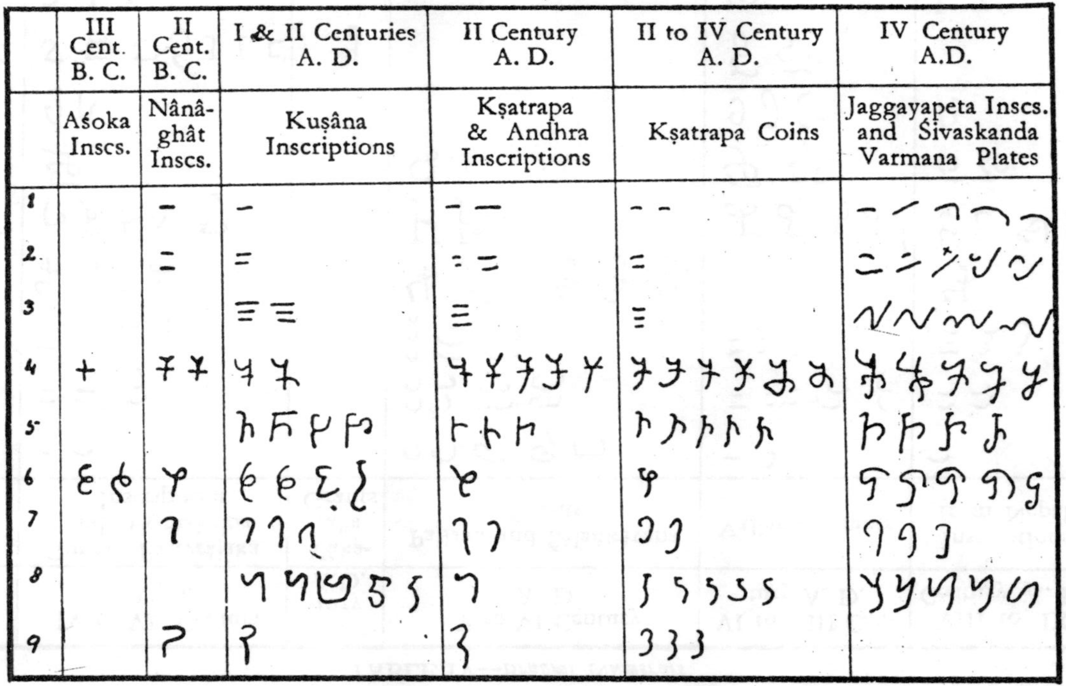

TABLE IV—*Brâhmî Numerals*

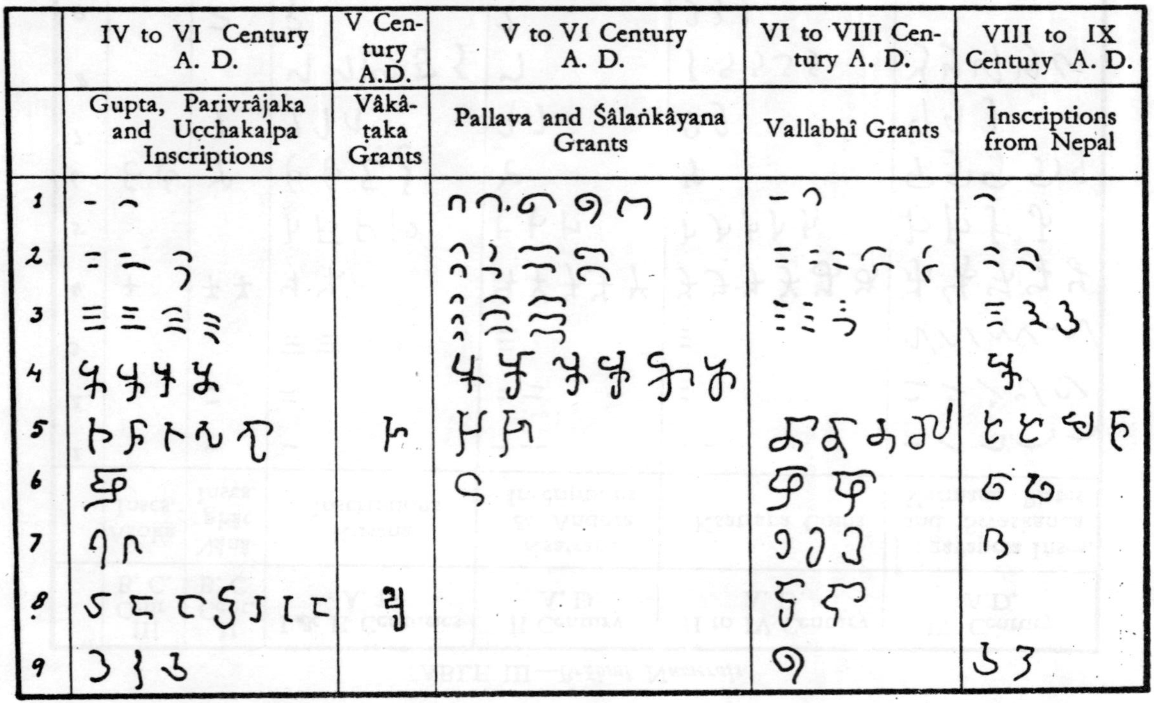

TABLE V—*Brâhmî Numerals*

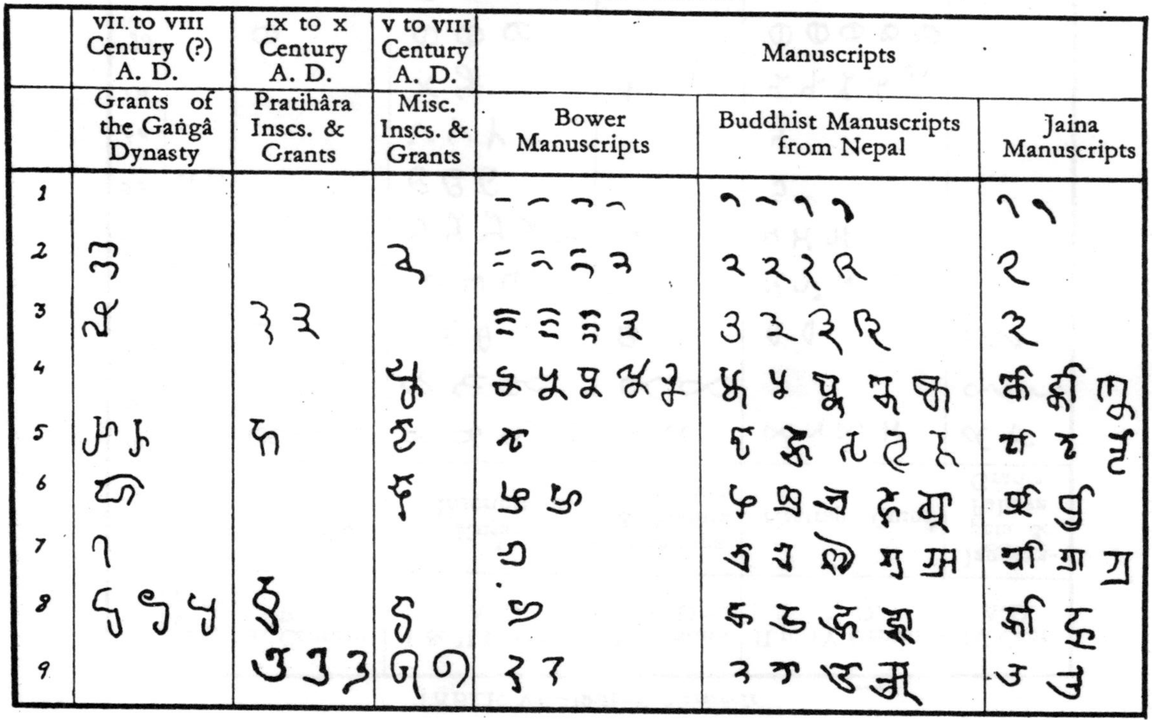

TABLE VI—*Brâhmî Numerals*

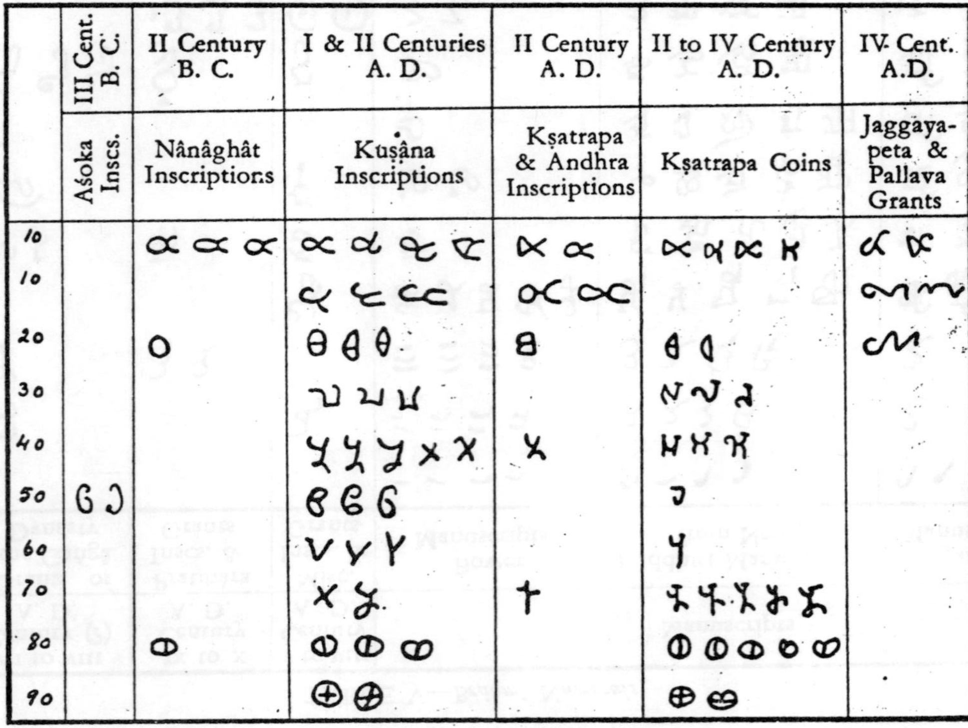

TABLE VII—*Brâhmî Numerals*

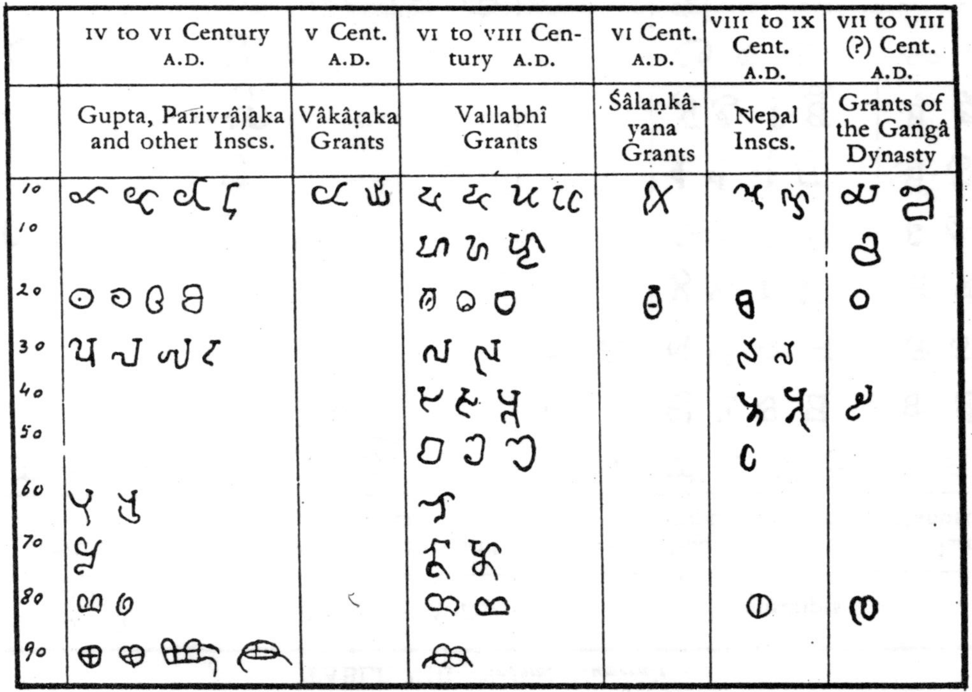

TABLE VIII—*Brâhmî Numerals*

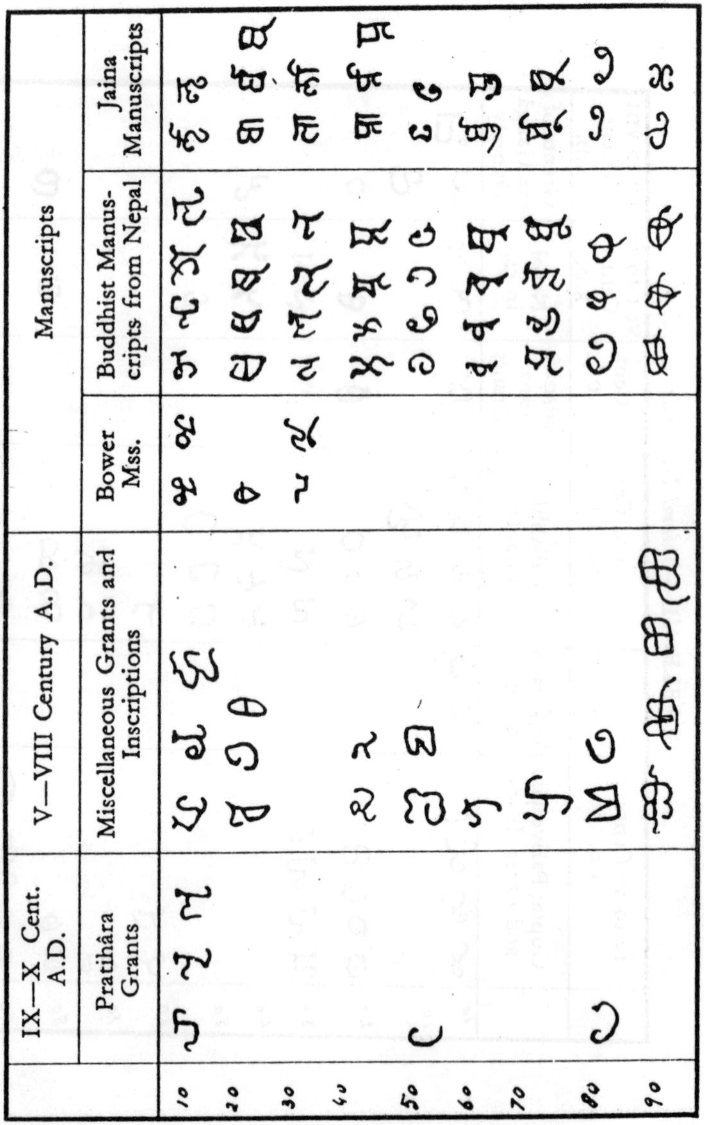

TABLE IX—*Brâhmî Numerals*

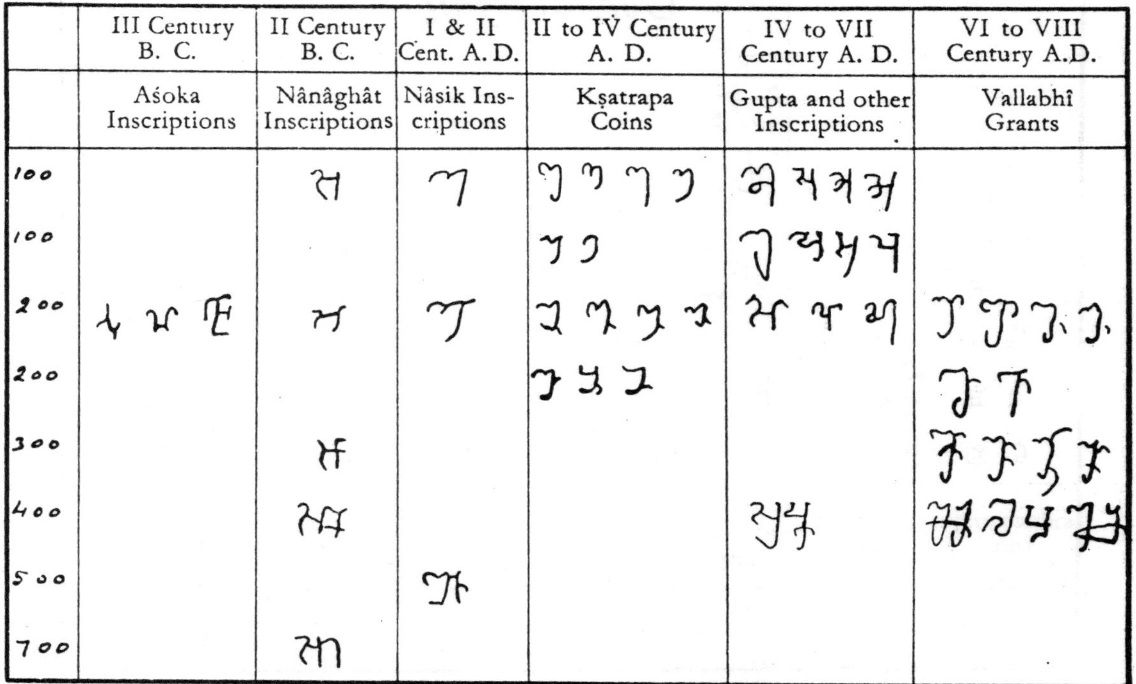

TABLE X—*Brâhmî Numerals*

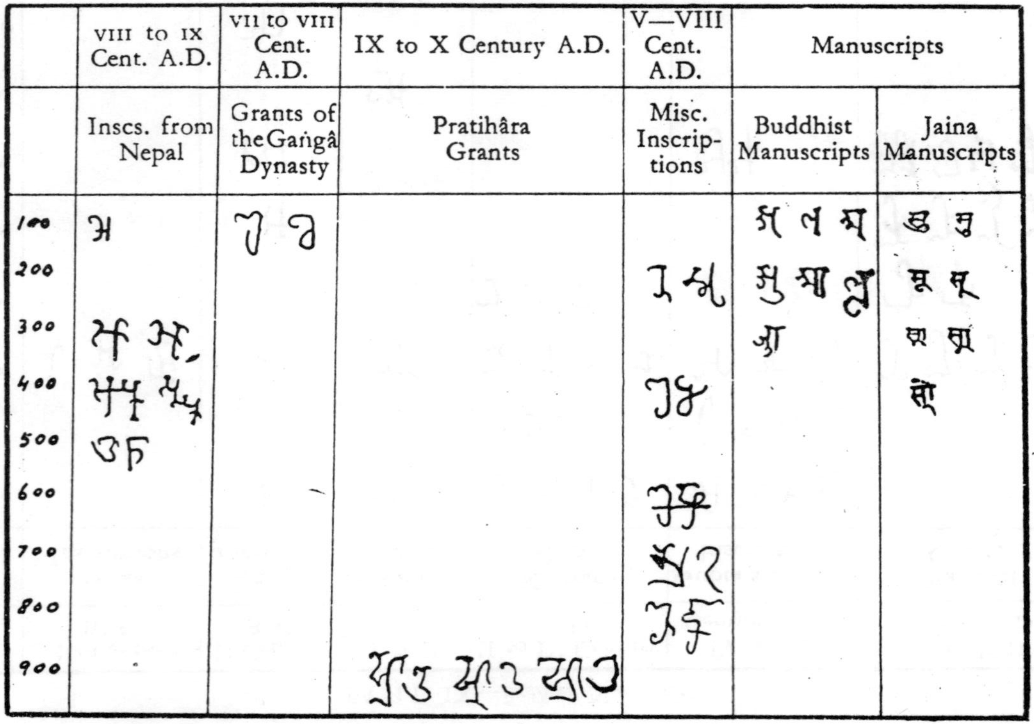

TABLE XI—*Brâhmî Numerals*

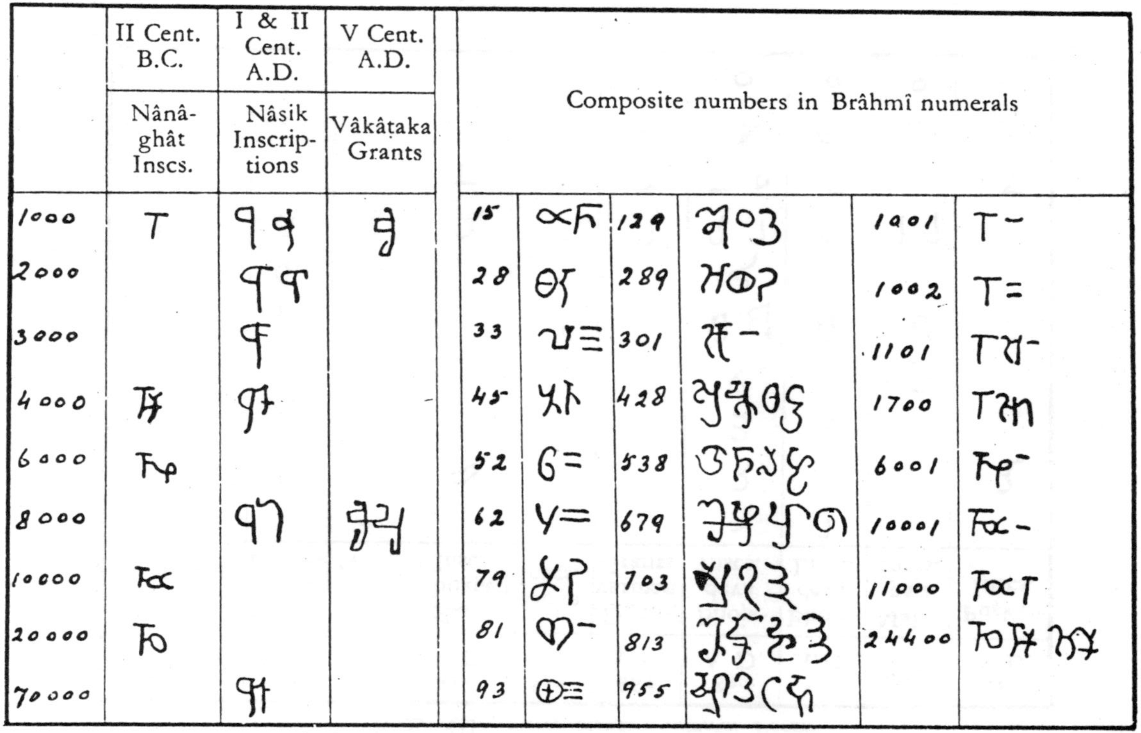

	II Cent. B.C. Nânâ-ghât Inscs.	I & II Cent. A.D. Nâsik Inscriptions	V Cent. A.D. Vâkâṭaka Grants	Composite numbers in Brâhmî numerals					
1000				15		129		1001	
2000				28		289		1002	
3000				33		301		1101	
4000				45		428		1700	
6000				52		538		6001	
8000				62		679		10001	
10000				79		703		11000	
20000				81		813		24400	
70000				93		955			

TABLE XII—*Early Hindu Numeral Forms*

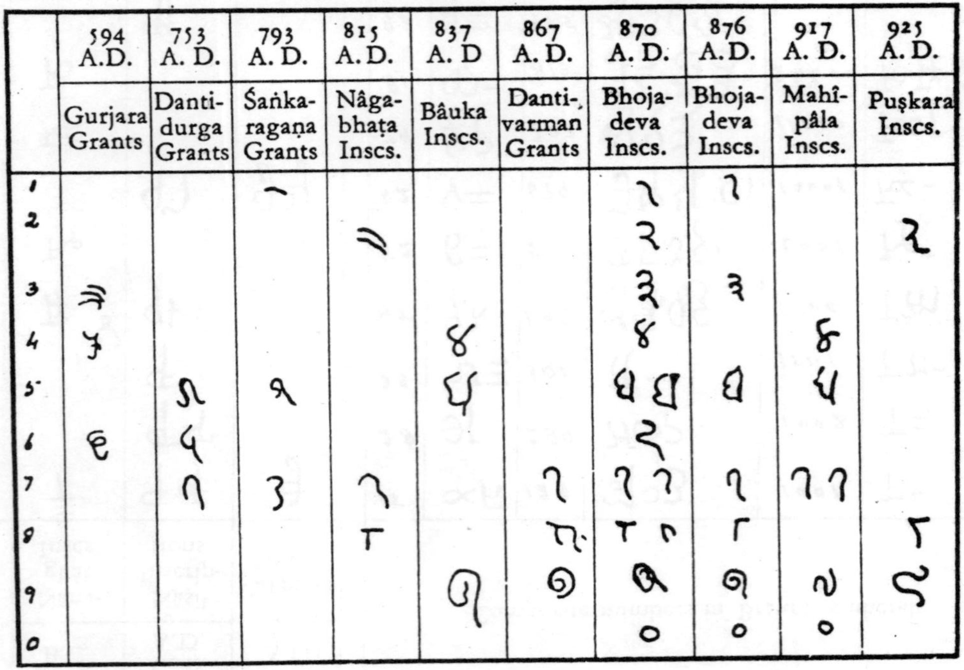

TABLE XIII—*Early Hindu Numeral Forms*

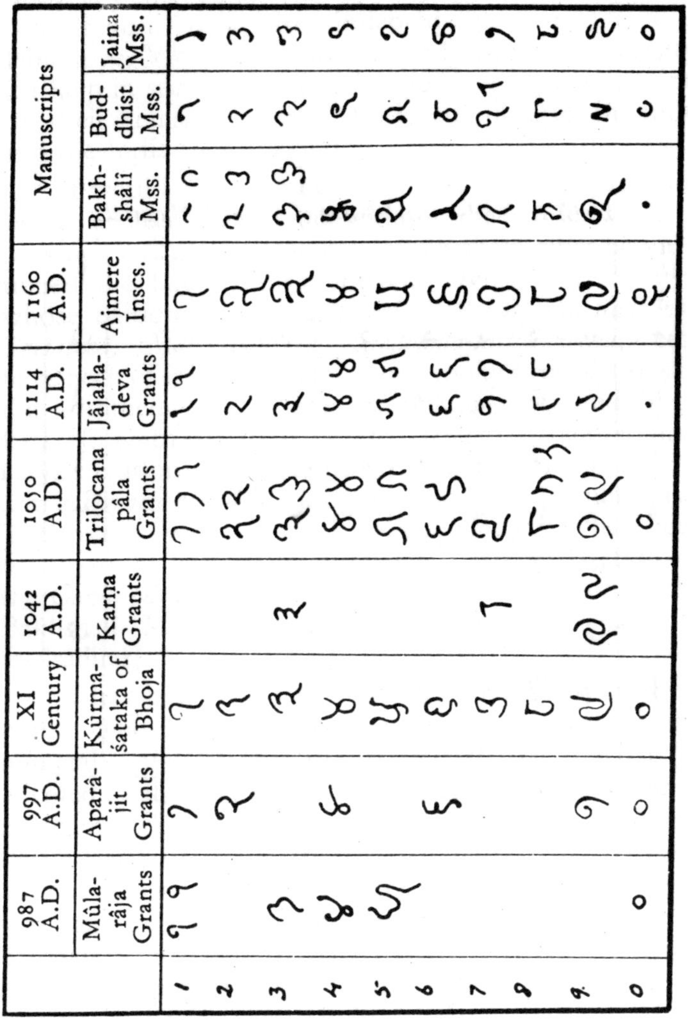

TABLE XIV—*Development of Nâgarî Numerals*

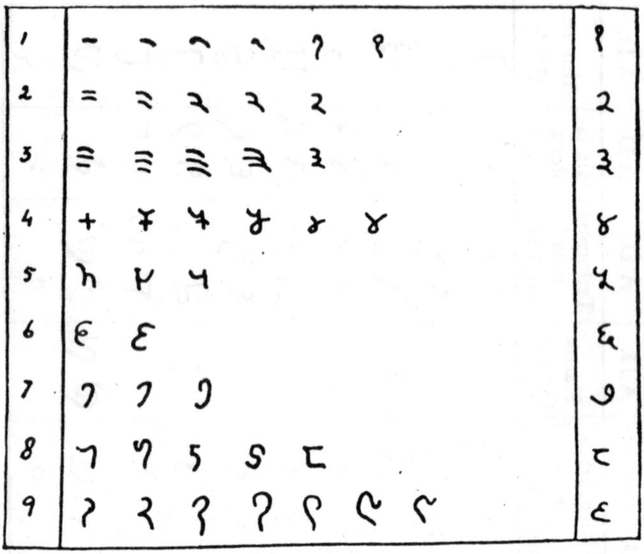

TABLE XV—*Numeral Forms in Modern Hindu Scripts*

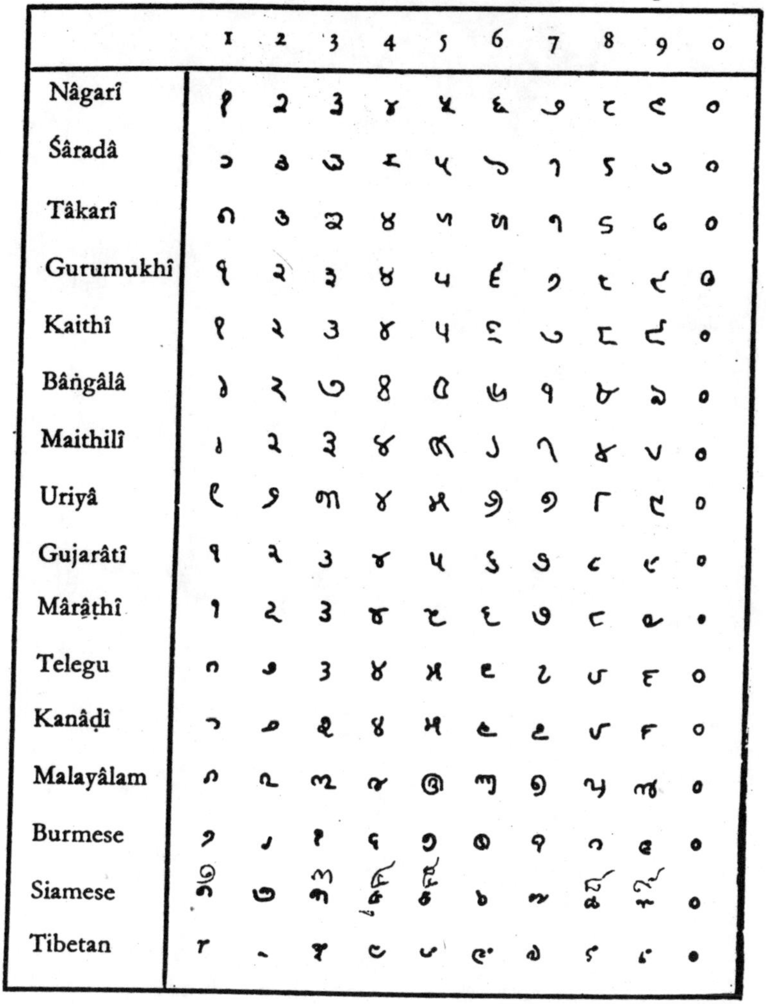

had not been understood before. In translating the texts we have tried to be as literal and faithful as possible without sacrificing the spirit of the original. Sometimes it has not been possible to find exact parallels to Sanskrit words and technical terms in English. In all such cases we have tried to maintain the spirit of the original in the English version.

The above plan of the book has been adopted in pursuance of our intention to place before those who have no access to the Sanskrit sources all evidence, unfavourable as well as favourable, so that they can judge for themselves the claims of Hindu mathematics, without depending solely on our statements. In order to facilitate comparison with the development of mathematics in other countries the various topics have been arranged generally in accordance with the sequence in Professor D. E. Smith's *History of Mathematics*, Vol. II. This has sometimes necessitated divergence from the arrangement of topics as found in the Hindu works on mathematics.

In search of material for the book we had to examine the literature of the Hindus, non-mathematical as well as mathematical, whether in Sanskrit or in Prākrit (Pālī and Ardha Māgadhī). Very few of the Hindu treatises on mathematics have been printed so far, and even these are not generally known. The manuscript works that exist in the various Sanskrit libraries in India and Europe are still less known. We have not spared labour in collecting as many of these as we could. Sanskrit mathematical works mentioned in the bibliography given at the end of this volume have been specially consulted by us. We are thankful to the authorities of the libraries at Madras, Bangalore, Trivandrum, Trippunithura and Benares, and those of the India Office (London) and the Asiatic Society of

CHAPTER II

ARITHMETIC

1. GENERAL SURVEY

Terminology and Scope. Arithmetic forms the major part of the Hindu works on *pâṭîgaṇita*. The word *pâṭîgaṇita* is a compound formed from the words *pâṭî*, meaning "board," and *gaṇita*, meaning "science of calculation;" hence it means the science of calculation which requires the use of writing material (the board).[1] It is believed that this term originated in a non-Sanskrit literature of India, a vernacular of Northern India. The oldest Sanskrit term for the board is *phalaka* or *paṭṭa*, not *pâṭî*. The word *pâṭî* seems to have entered into Sanskrit literature about the beginning of the seventh century A.D.[2] The carrying out of mathematical calculations was sometimes called *dhûlî-karma* ("dust-work"), because the figures were written on dust spread on a board or on the ground. Some later writers have used the term *vyakta-gaṇita* ("the science of calculation by the 'known'") for *pâṭîgaṇita* to distinguish it from algebra which was called *avyakta-gaṇita* ("the science of calculation by the 'unknown'"). The terms *pâṭîgaṇita* and *dhûlî-karma* were translated into Arabic when Sanskrit works were rendered into that language. The Arabic equivalents are *ilm-hisâb-al-takht* ("the science of

[1] Paper being scarce, a wooden board was generally used for making calculations even upto the 19th century.

[2] B. Datta, *American Math. Monthly*, XXXV, p. 526.

calculation on the board") and *hisâb-al-ghobâr* ("calculation on dust") respectively.

Bayley, Fleet and several others suspect that the origin of the term *pâṭî* in Hindu Mathematics lies in the use of the board as an abacus. This conjecture, however, is without foundation, as no trace of the use of any form of the abacus is found in India.

According to Brahmagupta[1] there are twenty operations and eight determinations in *pâṭîgaṇita*. He says:

"He who distinctly and severally knows the twenty logistics, addition, etc., and the eight determinations including (measurement by) shadow is a mathematician."

The twenty logistics, according to Pṛthudakasvâmî, are: (1) *saṁkalita* (addition), (2) *vyavakalita* or *vyutkalita* (subtraction), (3) *guṇana* (multiplication), (4) *bhâgahâra* (division), (5) *varga* (square), (6) *varga-mûla* (square-root), (7) *ghana* (cube), (8) *ghana-mûla* (cube-root), (9-13) *pañca jâti* (the five rules of reduction relating to the five standard forms of fractions), (14) *trairâśika* (the rule of three), (15) *vyasta-trairâśika* (the inverse rule of three), (16) *pañcarâśika* (the rule of five), (17) *saptarâśika* (the rule of seven), (18) *navarâśika* (the rule of nine), (19) *ekâdaśarâśika* (the rule of eleven), and (20) *bhâṇḍa-pratibhâṇḍa* (barter and exchange). The eight determinations are: (1) *miśraka* (mixture), (2) *śreḍhî* (progression or series), (3) *kṣetra* (plane figures), (4) *khâta* (excavation), (5) *citi* (stock), (6) *krâkacika* (saw), (7) *râśi* (mound), and (8) *châyâ* (shadow).

Of the operations named above, the first eight have been considered to be fundamental by Mahâvîra and later writers. The operations of duplation (doubling)

[1] *BrSpSi*, p. 172.

and mediation (halving), which were considered fundamental by the Egyptians, the Greeks and some Arab and western scholars, do not occur in the Hindu mathematical treatises. These operations were essential for those who did not know the place-value system of notation. They are not found in Hindu works, all of which use the place-value notation.

Sources. The only works available which deal exclusively with *pâṭîgaṇita* are: the *Bakhshâlî Manuscript* (*c*. 200), the *Triśatikâ* (*c*. 750), the *Gaṇita-sâra-saṁgraha* (*c*. 850), the *Gaṇita-tilaka* (1039), the *Lîlâvatî* (1150), the *Gaṇita-kaumudî* (1356), and the *Pâṭî-sâra* (1658). These works contain the twenty operations and the eight determinations mentioned above. Examples are also given to illustrate the use of the rules enunciated.

Besides these there are a number of astronomical works, known as *Siddhânta*, each of which contains a section dealing with mathematics. Âryabhaṭa I (499) was the first to include a section on mathematics in his *Siddhânta*, the *Âryabhaṭîya*. Brahmagupta (628) followed Âryabhaṭa in this respect, and after him it became the general fashion to include a section on mathematics in a *Siddhânta* work.[1] The earlier *Siddhânta* works do not possess this feature. The *Sûrya-siddhânta* (*c*. 300) does not contain a section on mathematics. The same is true of the *Vâsiṣṭha*, the *Pitâmaha* and the *Romaka* Siddhântas. Bhâskara I and Lalla,[2] although zealous followers of Âryabhaṭa I, did not emulate him in including a section on mathematics in their astronomical works.

[1] Amongst such works may be mentioned the *Mahâ-siddhânta* (950), the *Siddhânta-śekhara* (1036), the *Siddhânta-tattva-viveka* (1658), etc.

[2] It is stated by Bhâskara II that Lalla wrote a separate treatise on *pâṭîgaṇita*.

Exposition and Teaching. In India conciseness of composition, especially in scientific matters, was highly prized. The more compact and brief the composition, the greater was its value in the eyes of the learned. It is for this reason that the Indian treatises contain only a brief statement of the known formulæ and results, sometimes so concisely expressed as to be hardly understandable. This compactness is more pronounced in the older works; for instance, the exposition in the *Âryabhaṭîya* is more compact than in the later works.

This hankering after brevity, in early times, was due chiefly to the dearth of writing material, the fashion of the time and the method of instruction followed. The young student who wanted to learn *pâṭîgaṇita* was first made to commit to memory all the rules. Then he was made to apply the rules to the solution of problems (also committing the problems to memory). The calculations were made on a *pâṭî* on which dust was spread, the numbers being written on the dust with the tip of the fore-finger or by a wooden style, the figures not required being rubbed out as the calculation proceeded. Sometimes a piece of chalk or soap-stone was used to write on the *pâṭî*. Along with each step in the process of calculation the *sûtra* (rule) was repeated by the student, the teacher supervising and helping the student where he made mistakes. After the student had acquired sufficient proficiency in solving the problems contained in the text he was studying, the teacher set him other problems—a store of graded examples (probably constructed by himself or borrowed from other sources) being the stock-in-trade of every professional teacher. At this stage the student began to understand and appreciate the *rationale* of the easier rules. After this stage was reached the teacher gave proofs of the more difficult formulæ to the pupil.

It will be observed that the method of teaching pursued was extremely defective in so far as it was in the first two stages purely mechanical. A student who did not complete all the three stages knew practically nothing more than the mere mechanical application of a set of formulæ committed to memory; and as he did not know the *rationale* of the formulæ he was using, he was bound to commit mistakes in their application. It may be mentioned that not many teachers themselves could guide a pupil through all the stages of the teaching, and the earnest student, if he had a genuine desire to learn, had to go to some seat of learning or to some celeberated scholar to complete his training.

Mathematics is and has always been the most difficult subject to study, and as a knowledge of higher mathematics could not be turned to material gain there were very few who seriously undertook its study. In India, however, the religious practices of the Hindus required a certain amount of knowledge of astronomy and mathematics. Moreover, there have always been, from very early times, a class of people known as *gaṇaka* whose profession was fortune-telling. These people were astrologers, and in order to impress their clients with their learning, they used to have some knowledge of mathematics and astronomy. Thus it would appear that instruction in mathematics, upto a certain minimum standard, was available almost everywhere in India. As always happens, some of the pupils got interested in mathematics for its own sake, and took pains to make a thorough study of the subject and to add to it by writing commentaries or independent treatises.

Decay of Mathematics. All this was true when the times were normal. In abnormal times when there were foreign invasions, internal warfares or bad government and consequent insecurity, the study of mathematics and, in fact, of all sciences and arts languished. Al-Bîrûnî who visited north-western India after it had been in a very unsettled state due to recurrent Afghan invasions for the sake of plunder and loot complains that he could not find a *paṇḍit* who would explain to him the principles of Indian mathematics. Although Al-Bîrûnî's case was peculiar, for no respectable *paṇḍit* would agree to help a foreigner, especially one belonging to the same class as the invaders and the despoilers of temples, yet we are quite sure that in the Punjab there were very few good scholars at that time. We, however, know of at least one very distinguished mathematician, Śripati, who probably lived in Kashmir at that time.

It is certain, however, that after the 12th century very little original work was done in India. Commentaries on older works were written and some new works brought out, but none of these had sufficient merit as regards exposition or subject matter, so as to displace the works of Bhâskara II, which have held undisputed sway for nine centuries (as standard text books).

The Fundamental Operations. The eight fundamental operations of Hindu *gaṇita* are: (1) addition, (2) subtraction, (3) multiplication, (4) division, (5) square, (6) square-root, (7) cube and (8) cube-root. Most of these elementary processes have not been mentioned in the Siddhânta works. Âryabhaṭa I gives the rules for finding the square- and cube-roots only, whilst Brahmagupta gives the cube-root rule only. In the works on arithmetic (*pâṭîgaṇita*), the methods of addition and subtraction have not been mentioned at all or mentioned very briefly. Names of several methods of multiplication have been mentioned, but the methods themselves have been either very briefly described or not described at all. The modern method of division is briefly described in all the works and so are the methods of squaring, square-root, cubing and cube-root.

Although very brief descriptions of these fundamental operations are available, yet it is not difficult to reconstruct the actual procedure employed in performing these operations in ancient India. These methods have been well-known and taught to children, practically without any change, for the last fifteen hundred years or more. They are still performed in the old fashion on a *pâṭî* ("board") by those who have obtained their primary training in the Sanskrit pâṭhaśâlâ and not in the modern primary school. The details of these methods are also available to us in the various commentaries, viz., the commentary of Pṛthudakasvâmî and the several commentaries on Bhâskara's *Lîlâvatî*.

As already mentioned, the calculations were performed on sand spread on the ground (*dhûlî-karma*[1]) or on a *pâṭî* ("board"). Sometimes a piece of chalk or soap-stone (*pâṇḍu-lekha* or *śvetavarṇî*) was used to write on the *pâṭî*.[2] As the figures written were big, so several lines of figures could not be contained on the board. Consequently, the practice of obliterating figures not required for subsequent work was common. Instances of this would be found in the detailed method of working (the operations) given hereafter.

That all mathematical operations are variations of the two fundamental operations of addition and subtraction was recognised by the Hindu mathematicians

[1] Bhâskara II, *SiŚi*, *candragrahaṇâdhikâra*, 4.

[2] Bhâskara II: *khaṭikâyâ rekhâ ucchâdya...*, *i.e.*, "having drawn lines with a chalk...," quoted by S. Dvivedi in his *History of Mathematics* (in Hindi), Benares, 1910, p. 41.

from early times. Bhâskara I (*c.* 525) states:[1]

"All arithmetical operations resolve into two categories though usually considered to be four.[2] The two main categories are increase and decrease. Addition is increase and subtraction is decrease. These two varieties of operations permeate the whole of mathematics (*gaṇita*). So previous teachers have said: 'Multiplication and evolution are particular kinds of addition; and division and involution of subtraction. Indeed every mathematical operation will be recognised to consist of increase and decrease.' Hence the whole of this science should be known as consisting truly of these two only."

2. ADDITION

Terminology. Âryabhaṭa II (950) defines addition thus:

"The making into one of several numbers is addition."[3]

The Hindu name for addition is *saṁkalita* (made together). Other equivalent terms commonly used are *saṁkalana* (making together), *miśraṇa* (mixing), *sammelana* (mingling together), *prakṣepaṇa* (throwing together), *saṁyojana* (joining together), *ekîkaraṇa* (making into one), *yukti*, *yoga* (addition) and *abhyâsa*,[4] etc. The word *saṁkalita* has been used by some writers in the general sense of the sum of a series.[5]

The Operation. In all mathematical and astronomical works, a knowledge of the process of addition is

[1] The quotation is from his commentary on the *Âryabhaṭîya*.

[2] *i.e.*, addition, subtraction, multiplication and division.

[3] *MSi*, p. 143.

[4] This word has been used in the sense of addition in the *Śulba* only. It is used for multiplication in later works.

[5] *E.g.*, *Triś*, p.2; *GSS*, p. 17.

taken for granted. Very brief mention of it is made in some later works of elementary character. Thus Bhâskara II says in the *Lîlâvatî*:

"Add the figures in the same places in the direct or the inverse order."[1]

Direct Process. In the direct process of addition referred to above, the numbers to be added are written down, one below the other, just as at present, and a line is drawn at the bottom below which the sum is written. At first the sum of the numbers standing in the units place is written down, thus giving the first figure of the sum. The numbers in the tens place are then added together and their sum is added to the figure in the tens place of the partial sum standing below the line and the result substituted in its place. Thus the figure in the tens place of the sum is obtained; and so on. An alternative method used was to write the biggest addend at the top, and to write the digits of the sum by rubbing out corresponding digits of this addend.[2]

Inverse Process. In the inverse process, the numbers standing in the last place (extreme left) are added together and the result is placed below this last place. The numbers in the next place are then added and so on. The numbers of the partial sum are corrected, if necessary, when the figures in the next vertical line are added. For instance, if 12 be the sum of the numbers in the last place, 12 is put below the bottom line, 2 being directly below the numbers added; then, if the

[1] L, p. 2; direct (*krama*), *i.e.*, beginning from the units place; inverse (*utkrama*), *i.e.*, beginning from the last place on the left. The commentator Gaṅgâdhara says: *aṅkânâṃ vâmatogatiriti vitarkeṇa ekasthânâdi yojanaṃ kramaḥ utkramastu antyasthânâdi yojanaṃ*, *i.e.*, "According to the rule 'the numerals increase (in value) towards the left', the addition of units first is the direct method, the addition of figures in the last place first is the inverse method."

[2] Dvivedi, *History of Mathematics*, Benares, 1910, p. 60.

sum of the numbers in the next place is 13 (say), 3 is placed below the figures added and 1 is carried to the left. Thus the figure 2 of the partial sum 12 is rubbed out and substituted by 3.[1]

The Arabs used to separate the places by vertical lines, but this was not done by the Hindus.[2]

3. SUBTRACTION

Terminology. Âryabhaṭa II (950) gives the following definition of subtraction:

"The taking out (of some number) from the *sarvadhana* (total) is subtraction; what remains is called *śeṣa* (remainder)."[3]

The terms *vyutkalita* (made apart), *vyutkalana* (making apart), *śodhana* (clearing), *pâtana* (causing to fall), *viyoga* (separation), etc., have been used for subtraction. The terms *śeṣa* (residue) and *antara* (difference) have been used for the remainder. The minuend has been called *sarvadhana* or *viyojya* and the subtrahend *viyojaka*.

The Operation. Bhâskara II gives the method of subtraction thus:

"Subtract the numbers according to their places in the direct or inverse order."[4]

[1] The *Manorañjana* explains the process of addition thus: *Example*. Add 2, 5, 32, 193, 18, 10 and 100.

Sum of units	2,5,2,3,8,0,0	20
Sum of tens	3,9,1,1,0	14
Sum of hundreds	1,0,0,1	2
Sum of sums		360

The horizontal process has been adopted by the commentator so that both the 'direct' and 'inverse' processes may be exhibited by a single illustration. It was never used in practice.

[2] *Cf.* Taylor, *Lîlâwatî*, Bombay, 1816, Introduction, p. 14.

[3] *MSi*, p. 143.

[4] *L*, p. 2.

Direct Process. Sûryadâsa[1] explains the process of subtraction with reference to the example.

1000 — 360

thus:

"Hence making the subtraction as directed, six cannot be subtracted from the zero standing in the tens place, so taking ten and subtracting six from it, the remainder (four) is placed above (six), and this ten is to be subtracted from the next place. For, as the places of unit, etc., are multiples of ten, so the figure of the subtrahend that cannot be subtracted from the corresponding figure of the minuend is subtracted from ten, the remainder is taken and this ten is deducted from the next place. In this way this ten is taken to the last place until it is exhausted with the last figure. In other words, numbers upto nine occupy one place, the differentiation of places begins from ten, so it is known 'how many tens there are in a given number' and, therefore, the number that cannot be subtracted from its own place is subtracted from the next ten, and the remainder taken."

The above refers to the direct process, in which subtraction begins from the units place.

Inverse Process. The inverse process is similar, the only difference being that it begins from the last place of the minuend, and the previously obtained partial differences are corrected, if required. The process is suitable for working on a *pâṭî* (board) where figures can be easily rubbed out and corrected. This process seems to have been in general use in India, and was considered to be simpler than the direct process.[2]

[1] In his commentary on the *Lîlâvatî*.

[2] According to Gaṅgâdhara, the inverse process of working is easier in the case of subtraction, and the direct in the case of addition.

4. MULTIPLICATION

Terminology. The common Hindu name for multiplication is *guṇana*. This term appears to be the oldest as it occurs in Vedic literature. The terms *hanana*, *vadha*, *kṣaya*, etc. which mean "killing" or "destroying" have been also used for multiplication. These terms came into use after the invention of the new method of multiplication with the decimal place-value numerals; for in the new method the figures of the multiplicand were successively rubbed out (destroyed) and in their places were written the figures of the product.[1] Synonyms of *hanana* (killing) have been used by Ārya-bhaṭa I[2] (499), Brahmagupta (628), Śrīdhara (*c.* 750) and later writers. These terms appear also in the Bakhshâlî Manuscript.[3]

The term *abhyâsa* has been used both for addition and multiplication in the *Śulba* works (800 B.C.). This shows that at that early period, the process of multiplication was made to depend on that of repeated addition. The use of the word *parasparakṛtaṃ* (making together) for multiplication in the Bakhshâlî Manuscript[4] is evidently a relic of olden times. This ancient terminology proves that the definition of multiplication was "a process of addition resting on repetition of the multiplicand as many times as is the number of the multiplicator." This definition occurs in the commentary of the *Ārya-bhaṭîya* by Bhâskara I. The commentators of the *Lîlâvatî* give the same explanation of the method of multiplication.[5]

[1] See the *kapâṭa-sandhi* method of multiplication, pp. 138ff.

[2] *Ā*, ii. 19, 26, etc.

[3] *BMs*, 65 *verso*.

[4] *BMs*, 3 *verso*.

[5] Colebrooke, *Hindu Algebra*, p. 133.

The multiplicator was termed *guṇya* and the multiplier *guṇaka* or *guṇakâra*. The product was called *guṇana-phala* (result of multiplication) or *pratyutpanna* (lit. "reproduced," hence in arithmetic "reproduced by multiplication"). The above terms occur in all known Hindu works.

Methods of Multiplication. Âryabhaṭa I does not mention the common methods of multiplication, probably because they were too elementary and too well-known to be included in a Siddhânta work. Brahmagupta, however, in a supplement to the section on mathematics in his Siddhânta, gives the names of some methods with very brief descriptions of the processes:

"The multiplicand repeated, as in *gomûtrikâ*, as often as there are digits[1] in the multiplier, is severally multiplied by them and (the results) added (according to places); this gives the product. Or the multiplicand is repeated as many times as there are component parts[2] in the multiplier."[3]

"The multiplicand is multiplied by the sum or the difference of the multiplier and an assumed quantity and, from the result the product of the assumed quantity and the multiplicand is subtracted or added."[4]

Thus Brahmagupta mentions four methods: (1) *gomûtrikâ*, (2) *khaṇḍa*, (3) *bheda* and (4) *iṣṭa*. The common and well-known method of *kapâṭa-sandhi* has been omitted by him.

[1] *khaṇḍa*, translated as "integrant portions" by Colebrooke.

[2] *bheda*, i.e., portions which added together make the whole, or aliquot parts which multiplied together make the entire quantity.

[3] *BrSpSi*, p. 209; Colebrooke, *l.c.*, p. 319.

[4] *BrSpSi*, p. 209. Colebrooke (*l.c.*, p. 320) thinks that this is a method to obtain the true product when the multiplier has been taken to be too great or too small by mistake. This view is incorrect.

Śrîdhara mentions four methods of multiplication: (1) *kapâṭa-sandhi*, (2) *tastha*, (3) *rûpa-vibhâga* and (4) *sthâna-vibhâga*. Mahâvîra mentions the same four. Âryabhaṭa II mentions only the common method of *kapâṭa-sandhi*. Bhâskara II, besides the above four, mentions Brahmagupta's method of *iṣṭa-guṇana*. The five methods given by Bhâskara II were mentioned earlier by Śrîpati in the *Siddhânta-śekhara*. Gaṇeśa[1] (1545) mentions the gelosia method of multiplication under the name of *kapâṭa-sandhi* and adds that the intelligent can devise many more methods of multiplication. The method is also given in the *Gaṇita-mañjarî*. We have designated it as *kapâṭa-sandhi* (*b*).

Seven[2] distinct modes of multiplication employed by the Hindus are given below. Some of these are as old as 200 A.D. These methods were transmitted to Arabia in the eighth century and were thence communicated to Europe, where they occur in the writings of mediæval mathematicians.

Door-junction Method. The Sanskrit term for the method is *kapâṭa-sandhi*. Śrîdhara[3] describes it thus:

"Placing the multiplicand below the multiplier as in *kapâṭa-sandhi*,[4] multiply successively, in the direct or inverse order, moving the multiplier each time. This method is called *kapâṭa-sandhi*."

Âryabhaṭa II[5] (950) gives the following without name:

[1] Commentary on the *Lîlâvatî*, MSS No. I. B. 6. in the Asiatic Soc. of Bengal, Calcutta, pp. 17, 18. In this work only two methods are given, (1) *kapâṭa-sandhi* and (2) *kapâṭa-sandhi* (*b*).

[2] Or ten if we count also the sub-divisions under each head.

[3] *Triś*, pp. 3f.

[4] *kapâṭa* means "door" and *sandhi* means "junction"; hence *kapâṭa-sandhi* means "the junction of doors."

[5] *MSi*, p. 143; the inverse method only has been given.

"Place the first figure of the multiplier over the last figure of the multiplicand, and then multiply successively all the figures of the multiplier by each figure of the multiplicand."

Śrîpati[1] (1039) gives the name *kapâṭa-sandhi* and states:

"Placing the multiplicand below the multiplier as in the junction of two doors multiply successively (the figures of the multiplicand) by moving it (the multiplier) in the direct or inverse order."

Mahâvîra refers to a method known as *kapâṭa-sandhi*, but does not give the details of the process.[2] Bhâskara II gives the method but not the name, while Nârâyaṇa (1356) gives the method in almost the same words as Śrîdhara, and calls it *kapâṭa-sandhi*.

The main features of the method are (*i*) the relative positions of the multiplicand and the multiplier and (*ii*) the rubbing out of figures of the multiplicand and the substitution in their places of the figures of the product. The method owes its name *kapâṭa-sandhi* to the first feature, and the later Hindu terms meaning "killing" or "destroying" for multiplication owe their origin to the second feature. The occurrence of the terms *hanana*, *vadha*, etc., in the works of Âryabhaṭa I and Brahmagupta, and in the Bakhshâlî Manuscript show beyond doubt that this method was known in India about 200 A.D.

The following illustrations[3] explain the two processes of multiplication according to the *kapâṭa-sandhi* plan:

[1] *SiŚe*, xiii. 2; *GT*, 15.

[2] *GSS*, p. 9.

[3] The illustrations are based on the accounts given in the commentaries on the *Lîlâvatî*, especially the *Manorañjana* which gives more details.

Direct Process: This method of working does not appear to have been popular. It has not been mentioned by writers after the 11th century, Śrîpati (1039) being the last writer to mention it.

Example. To multiply 135 by 12.

The numbers are written down on the *pâṭî* thus:

12
135

The first digit of the multiplicand (5) is taken and multiplied with the digits of the multiplier. Thus $5\times 2=10$; 0 is written below 2, and 1 is to be carried over.[1] Then $5\times 1=5$; adding 1 (carried over), we get 6. 5 which is no longer required is rubbed out and 6 written in its place. Thus we have

12
1360

The multiplier is then moved one place towards the left, and we have

12
1360

Now, 12 is multiplied by 3. The details are: $3\times 2=6$; this 6 added to the figure 6 below 2 gives 12. 6 is rubbed out and 2 substituted in its place. 1 is carried over. Then $3\times 1=3$; 3 plus 1 (carried over)$=4$. 3 is rubbed out and 4 substituted. After the multiplier 12 has been moved another place towards the left, the figures on the *pâṭî* stand thus:

12
1420

Then, $1\times 2=2$; $2+4=6$; 4 is rubbed out and 6 substituted. $1\times 1=1$, which is placed to the left of 6.

[1] For this purpose it was probably noted in a separate portion of the *pâṭî* by the beginner.

As the operation has ended, 12 is rubbed out and the *pâṭî* has

1620

Thus the numbers 12 and 135 have been *killed*[1] and a new number 1620 is born (*pratyutpanna*).[2]

The reader will note that the position of the multiplier and its motion serve two important purposes, *viz.*, (*i*) the last figure of the multiplier indicates the digit of the multiplicand by which multiplication is to be performed and, (*ii*) the product is to be added to the number standing underneath the digit of the multiplier multiplied.

Sometimes the product of a digit of the multiplicand and the multiplier extends beyond the last place of the multiplier. In such cases, the last figure of the partial product is noted separately. The reader should note this fact in the case, 135×99, by performing the operation according to the above process.

The beginner was liable to commit mistakes in such cases, (*i*) of not correctly taking into account the separately noted number, or (*ii*) of rubbing out the digit of the multiplicand beyond the last digit of the multiplier. For these reasons, this process was not in general use and the inverse process was preferred.

Inverse Method: There appear to have been two varieties of the inverse method.

(*a*) In the first the numbers are written thus:

12
135

Multiplication begins with the last digit of the multiplicand. Thus 1×2=2; 1 is rubbed out and 2 substi-

[1] This explains the use of the term *hanana* (killing) and its synonyms for multiplication.

[2] Hence the product was termed *pratyutpanna*.

tuted; then $1 \times 1 = 1$, this is written to the left;[1] the multiplier 12 is moved to the next figure. The work on the *pâṭî* stands thus:

$$\begin{array}{r} 12 \\ \cdot\ \ 1235 \end{array}$$

Then, $3 \times 2 = 6$; 3 is rubbed out and 6 substituted; then $3 \times 1 = 3$ and $3 + 2 = 5$; 2 is rubbed out and 5 substituted in its place. The multiplier having been moved, the work on the *pâṭî* stands thus:

$$\begin{array}{r} 12 \\ 1565 \end{array}$$

Now, $5 \times 2 = 10$; 5 is rubbed out and 0 substituted in its place; then $5 \times 1 = 5$; $5 + 1 = 6$; $6 + 6 = 12$; 6 is rubbed out and 2 substituted, and 1 is carried over; then $1 + 5 = 6$, 5 is rubbed out and 6 substituted in its place. The *pâṭî* has now,

$$1620$$

as the product (*pratyutpanna*). The figures to be carried over are noted down on a separate portion of the *pâṭî* and rubbed out after addition.

(*b*) In the second the partial multiplications (*i.e.*, the multiplications by the digits of the multiplicand) are carried out in the direct manner. These partial multiplications, however, seem to have been carried out in the inverse way, this being the general fashion. The following example will illustrate the method of working:

Example. Multiply 324 by 753

The multiplier and the multiplicand are arranged thus:

$$\begin{array}{c} 753 \\ 324 \end{array}$$

[1] Or the alternative plan: $1 \times 1 = 1$ and then $1 \times 2 = 2$, thus giving 12 in the place of 1 in the multiplicand, etc.

Multiplication begins with the last place of the multiplier. 3×7 gives 21; 1 is placed below the 7 of the multiplier and 2 to its left, thus:

753
21 324

Then 3×5 gives 15; 5 is placed below the 5 of the multiplier and 1 carried to the left; the 1 obtained in the previous step is rubbed out and (1+1)=2 is substituted, giving

753
225324

Then 3×3 gives 9; the 3 of the multiplicand is rubbed out and 9 substituted. The work on the *pâṭî* now stands thus:

753
225924

The multiplier is now moved one place to the right giving

753
225924

Then multiplying 7 by 2 we get 14. This 14 being set below the 7 gives

753
239924

Multiplying 5 by 2 and setting the result below it, we obtain

753
240924

Finally multiplying 3 by 2 and rubbing out 2, which is required no longer, and substituting 6 in its place, we get

753
240964

The multiplier is then moved one step further giving

753
240964

Multiplying by 4 the digits of the multiplier 753, and setting the results as before we obtain

(*i*)	753 243764	multiplying 7×4 and setting the result;
(*ii*)	753 243964	multiplying 5×4 and setting the result;
(*iii*)	753 243972	multiplying 3×4 and setting the result.

It may be again remarked that the position and motion of the multiplier play a very important part in the above process. The digits of the multiplier are also successively rubbed out in order to avoid confusion, thus 7 is rubbed out at stage (*i*), 5 at (*ii*) and 3 at (*iii*).

The following variation of the above process is also found:[1]

"Multiplicand 135, multiplier 12; the multiplier placed at the last place of the multiplicand gives

12
135

According to the rule 'the numerals progress to the left' the last figure of the multiplicand (the figure 1) is multiplied by 12. Then after moving (12) we get

12
1235

Again, the figure 3 next to the last of the multiplicand being multiplied by the multiplier 12 gives

12
1265
3

[1] *Lîlâvatyudâharaṇa* by Kṛpârâma Daivajña, Asiatic Society of Bengal, Calcutta, Ms. No. III. F. 110. A.

Then after moving (12) we get

12
1265
3

Again, multiplying the first figure 5 of the multiplicand with the multiplier 12, we get

12
1260
36

Then rubbing out the multiplier, the numbers

1260
36

being added according to places give 1620."

Transmission to the West. The *kapâṭa-sandhi* method of multiplication was transmitted to the Arabs who learnt the decimal arithmetic from the Hindus. It occurs in the works of Al-Khowârîzmî (825), Al-Nasavî[1] (*c.* 1025) Al-Haṣṣâr[2] (*c.* 1175), Al-Kalasâdî[3] (*c.* 1475) and many others. The following illustration is taken from the work of Al-Nasavî who calls this method *al-amal al-hindî* and *târik al-hindî* ("the method of the Hindus"):

Example. To multiply 324×753

```
 43
 3̸0̸9
 2̸9̸7̸7
21̸5̸9̸6̸2   ∴ Product = 243972.
   324
 753
  753
   753
```

[1] F. Woepeke, I (6), p. 407.
[2] H. Suter, *Bibl. Math.*, II (3), p. 16.
[3] *Ibid*, p. 17.

In the above the arrangement of the multiplicand and multiplier is just the same as in the Hindu method. The multiplier is moved in the same way. As the work is performed on paper, the figures are crossed out instead of being rubbed out.

It may be mentioned that in Europe, the method is found reproduced in the work of Maximus Planudes.

Gelosia Method. The method known as the 'gelosia',[1] has been described in the *Gaṇita-mañjarî* (16th century) as the *kapâṭa-sandhi* method. It appears also in Gaṇeśa's commentary on the *Lîlâvatî*. As the description of the *kapâṭa-sandhi* given by the older mathematicians is incomplete and sketchy, it is difficult to say whether Gaṇeśa is right in identifying the gelosia method with the *kapâṭa-sandhi* of older writers. In our opinion Gaṇeśa's identification is incorrect.[2]

We are at present unable to say definitely whether this method is a Hindu invention or was borrowed from the Arabs who are said to have used it in the 13th century.[3] It occurs in some Arab works of the 14th century, and also in Europe about the same time. Gaṇeśa was undoubtedly one of the best mathematicians of his time and the fact that he identified this method with the *kapâṭa-sandhi* which is the oldest known method shows that the gelosia method must have been in use in India from a long time before him.

The only available description of the method runs as follows:

"(Construct) as many compartments as there are places in the multiplicand and below these as many

[1] We shall designate it as *kapâṭa-sandhi* (*b*) method.
[2] *Cf.* the quotation from Śrîpati given before, p. 137.
[3] Smith, *History*, II, p. 115.

as there are places in the multiplier; the oblique lines in the first, in the one below, and in the other (compartments) are produced. Multiply each place of the multiplicand, by the places of the multiplier (which are) one below the other and set the results in the compartments. The sum taken obliquely on both sides of the oblique lines in the compartments gives the product. This is the *kapâṭa-sandhi*."[1]

The following illustration is taken from Gaṇeśa's commentary on the *Lîlâvatî*:

To Multiply 135 by 12

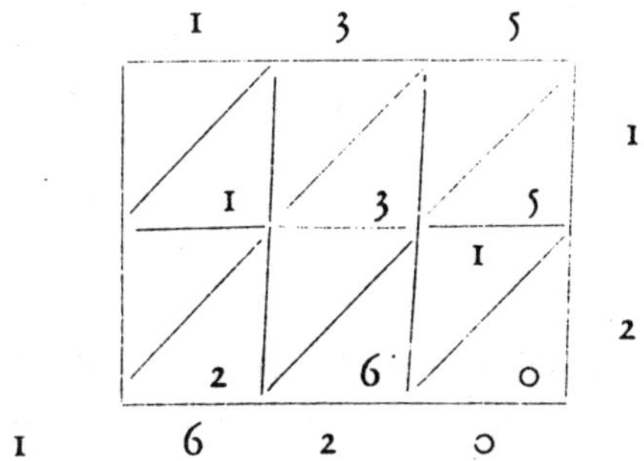

Cross Multiplication Method. This method has been mentioned by Śrîdhara, Mahâvîra, Śrîpati and some later writers as the *tastha* method. These writers, however, do not explain the method. Śrîdhara simply states: "The next (method) in which (the multiplier) is stationary is the *tastha*."[2] The method is algebraic and has been compared to *tiryak-guṇana* or *vajrâbhyâsa* (cross multiplication) used in algebra.[3] It has been explained by Gaṇeśa (*c.* 1545) thus:

[1] Translated from the *Gaṇita-mañjarî* of Gaṇeśa, son of Ḍhuṇḍhirâja.

[2] *Triś*, p. 3.

[3] Colebrooke, *l.c.*, p. 171, fn. 5.

"That method of multiplication in which the numbers stand in the same place,[1] is called *tastha-guṇana*. It is as follows: after setting the multiplier under the multiplicand multiply unit by unit and note the result underneath. Then as in *vajrâbhyâsa* multiply unit by ten and ten by unit, add together and set down the result in the line. Next multiply unit by hundred, hundred by unit and ten by ten, add together and set down the result as before; and so on with the rest of the digits. This being done, the line of results is the product."[2]

This method was known to the Hindu scholars of the 8th century, or earlier. The method seems to have travelled to Arabia and thence was transmitted to Europe, where it occurs in Paciôli's *Suma*[3] and is stated to be "more fantastic and ingenious than the others." Gaṇeśa has also remarked that "this (method) is very fantastic and cannot be learnt by the dull without the traditional oral instructions."

Multiplication by Separation of Places. This method of multiplication known as *sthâna-khaṇḍa*, is based on the separation of the digits of the multiplicand or of the multiplier. It has been mentioned in all the works from 628 A.D. onwards. Bhâskara II describes the method as follows:

"Multiply separately by the places of figures and add together."[4]

With reference to the example 135×12, Bhâskara II explains the method thus:

[1] In contra-distinction to the method in which the multiplier moves from one place to another.

[2] Gaṇeśa's commentary on the *Lîlâvatî*, i, 4-6.

[3] Smith (*l.c.*, II, p. 112) quotes from this work.

[4] *L*, p. 3.

"Taking the digits separately, *viz.*, 1 and 2, the multiplicand being multiplied by them severally, and the products added together according to places, the result is 1620."

Various arrangements appear to have been employed for writing down the working. Some of these are given below:

(*i*)[1]

```
  135
 12
 ----
 12
  36
   60
 ----
 1620
```

(*ii*)[2]

```
12 12 12
 1  3  5
--------
 1260
  36
--------
 1620
```

(*iii*)[3]

```
135  135
  1    2
--------
   270
  135
--------
  1620
```

Zigzag Method. The method is called *gomûtrikâ.*[4] It has been described by Brahmagupta. It is in all

[1] In a manuscript used by Taylor, see his *Lîlâwatî*, pp. 8-9.

[2] This arrangement is found in the commentary of Gaṅgâdhara on the *Lîlâvatî*, in the library of the Asiatic Society of Bengal, Calcutta.

[3] Found in Gaṅgâdhara, *l.c.*

[4] The word *gomûtrikâ*, means "similar to the course of cow's urine," hence "zigzag." Colebrooke's reading *gosûtrikâ* is incorrect. The method of multiplication of astronomical quantities is called *gomûtrikâ* even upto the present day by the paṇḍits.

essentials the same as the *sthâna-khaṇḍa* method. The following illustration is based on the commentary of Pṛthudakasvâmî.

Example. To multiply 1223 by 235

The numbers are written thus:

```
2   1223
3    1223
5     1223
```

The first line of figures is then multiplied by 2, the process beginning at the units place, thus: $2 \times 3 = 6$; 3 is rubbed out and 6 substituted in its place, and so on. After all the horizontal lines have been multiplied by the corresponding numbers on the left in the vertical line, the numbers on the *pâṭî* stand thus:

```
2 4 4 6
  3 6 6 9
    6 1 1 5
-----------
2 8 7 4 0 5
```

after being added together as in the present method.

The *sthâna-khaṇḍa* and the *gomûtrikâ* methods resemble the modern plan of multiplication most closely. The *sthâna-khaṇḍa* method was employed when working on paper.

Parts Multiplication Method. This method is mentioned in all the Hindu works from 628 A.D. onwards. Two methods come under this head:

(*i*) The multiplier is broken up into two or more parts whose sum is equal to it. The multiplicand is then multiplied severally by these and the results added.[1]

(*ii*) The multiplier is broken up into two or more aliquot parts. The multiplicand is then multiplied by

[1] Thus $12 \times 135 = (4+8) \times 135 = (4 \times 135) + (8 \times 135)$.

one of these, the resulting product by the second and so on till all the parts are exhausted. The ultimate product is the result.[1]

These methods are found among the Arabs and the Italians, having been obtained from the Hindus. They were known as the "Scapezzo" and "Repiego" methods respectively among the Italians.[2]

Algebraic Method. This method was known as *iṣṭa-guṇana*. Brahmagupta's description of the method has been already quoted. Bhâskara II explains it thus:

"Multiply by the multiplicator diminished or increased by an assumed number, adding or subtracting (respectively) the product of the multiplicand and the assumed number."[3]

This is of two kinds according as we (*i*) add or (*ii*) subtract an assumed number. The assumed number is so chosen as to give two numbers with which multiplication will be easier than with the original multiplier. The two ways are illustrated below:

(*i*) $$135 \times 12 = 135 \times (12+8) - 135 \times 8$$
$$= 2700 - 1080 = 1620$$

(*ii*) $$135 \times 12 = 135 \times (12-2) + 135 \times 2$$
$$= 1350 + 270 = 1620$$

This method was in use among the Arabs[4] and in Europe[5], obviously under Hindu influence.

[1] Thus $12 \times 135 = 3 \times 135 \times 4$.

[2] Smith, *History*, II, p. 117.

[3] *L*, p. 3.

[4] *E.g.*, Behâ Eddin (*c.* 1600). See G. Eneström, *Bibl. Math.*, VII (3), p. 95.

[5] *E.g.*, Widman (1489), Riese (1522), etc. See Smith, *l.c.*, p. 120.

5. DIVISION

Terminology. Division seems to have been regarded as the inverse of multiplication. The common Hindu names for the operation are *bhâgahâra*, *bhâjana*, *haraṇa*, *chedana*, etc. All these terms literally mean "to break into parts," *i.e.*, "to divide," excepting *haraṇa* which denotes "to take away." This term shows the relation of division to subtraction. The dividend is termed *bhâjya*, *hârya*, etc., the divisor *bhâjaka*, *bhâgahara* or simply *hara*, and the quotient *labdhi* "what is obtained" or *labdha*.

The Operation. Division was considered to be a difficult and tedious operation by European scholars even as late as the 15th and 16th centuries;[1] but in India the operation was not considered to be difficult, as the most satisfactory method of performing it had been evolved at a very early period. In fact, no Hindu mathematician seems to have attached any great importance to this operation. Âryabhaṭa I does not mention the method of division in his work. But as he has given the modern methods for extracting square- and cube-roots, which depend on division,[2] we conclude that the method of division was well-known in his time and was not described in the *Âryabhaṭîya* as it was considered to be too elementary. Most Siddhânta writers have followed Âryabhaṭa in excluding the process of division from their works, *e.g.*, Brahmagupta (628), Śrîpati (1039), and some others.

A method of division by removing common factors seems to have been employed in India before the invention of the modern plan. This removal of common

[1] Smith, *l.c.*, p. 132.
[2] He has used the technical term *labdha* for the quotient.

factors is mentioned in early Jaina works.[1] It has been mentioned by Mahâvîra who knew the modern method, probably because it was considered to be suitable in certain particular cases:

"Putting down the dividend and below it the divisor, and then, having performed division by the method of removing common factors, give out the resulting (quotient)."[2]

The modern method of division is not found in the Bakhshâlî Manuscript, although the name of the operation is found at several places. The absence of the method may be due to the mutilated form of the text, although it is quite possible that the method was not known at that early period (200 A.D.).

The Method of Long Division. The modern method of division is explained in the works on *pâṭîgaṇita*, the earliest of which, Śrîdhara's *Triśatikâ*, gives the method as follows:[3]

"Having removed the common factor, if any, from the divisor and the dividend, divide by the divisor (the digits of the dividend) one after another in the inverse[4] order."

Mahâvîra says:[5]

"The dividend should be divided by the divisor (which is) placed below it, in the inverse order, after having performed on them the operation of removing common factors."

[1] *Tatvârthâdhigama-sûtra*, *Bhâṣya* of Umâsvâti (*c*. 160, ed. by H. R. Kapadia, Bombay, 1926, Part I, ii. 52, p. 225.

[2] *GSS*, p. 11. The method would not give the quotient unless the dividend be completely divisible by the divisor.

[3] *Triś*, p. 4.

[4] *Pratiloma*.

[5] *GSS*, p. 11; *cf.* Rangacarya's translation.

Âryabhaṭa II gives more details of the process:[1]

"Perform division having placed the divisor below the dividend; subtract from (the last digits of the dividend) the proper multiple of the divisior; this (the multiple) is the partial quotient, then moving the divisor divide what remains, and so on."

Bhâskara II,[2] Nârâyaṇa[3] and others give the same method.

The following example will serve to illustrate the Hindu method of performing the operation on a *pâṭî*:

Example. Divide 1620 by 12.

The divisor 12 is placed below the dividend thus :

1620
12

The process begins from the extreme left of the dividend, in this case the figure 16. This 16 is divided by 12. The quotient 1 is placed in a separate line, and 16 is rubbed out and the remainder 4 is substituted in its place. The subtraction is made by rubbing out figures successively as each figure of the product to be subtracted is obtained. Thus, the partial quotient 1, being written, the procedure is

	1
1620	line of quotients
12	

$1\times 1=1$, so 1 of the dividend is rubbed out (as $1-1=0$); then $1\times 2=2$, so 4 is substituted in the place of 6 (as $6-2=4$). The figures on the *pâṭî* are:

	1
420	line of quotients
12	

[1] *MSi*, p. 144.

[2] Bhâskara gives the process briefly as follows: "That number, by which the divisor being multiplied, balances the last digit of the dividend gives the (partial) quotient, and so on." (*L*, p. 3)

[3] *GK*, i. 16.

The divisor 12 is now moved one place to the right giving

420	1
12	line of quotients

42 is then divided by 12. The resulting quotient 3 is set in the "line of quotients," 42 is rubbed out and the remainder 6 substituted in its place. The figures now stand thus:

60	13
12	line of quotients

Moving the divisor one place to the right, we have

60
12

On division being performed, as before the resulting quotient 5 is set in the "line of quotients" and 60 is rubbed out leaving no remainder. The line of quotients[1] has

135

which is the required result.

The above process, when the figures are not obliterated and the successive steps are written down one below the other, becomes the modern method of long division.

The method seems to have been invented in India about the 4th century A.D., if not earlier. It was transmitted to the Arabs, where it occurs in Arabic works from the 9th century onwards.[2] From Arabia the method travelled to Europe where it came to be known as the galley (*galea*, *batello*) method.[3] In this variation

[1] The "line of quotients" was usually written above the dividend.

[2] Al-Khowârîzmî (*c.* 825), Al-Nasavî (*c.* 1025); cf. Smith, *l.c.*, pp. 138-139.

[3] Also called the 'scratch method'.

of the method, the figures obtained at successive stages are written and crossed out, for the work is carried out on paper (where the figures cannot be rubbed out). The method was very popular in Europe from the 15th to the 18th century.[1] The above example worked on the galley plan would be represented thus:

	4	
I	~~1~~~~6~~20	1
	~~1~~~~2~~2	
	1	
II	~~1~~	
	~~4~~6	
	~~1~~~~6~~~~2~~0	13
	~~1~~~~2~~~~2~~2	
	~~1~~1	
III	~~1~~~~1~~	
	~~4~~~~6~~	
	~~1~~~~6~~~~2~~~~0~~	135
	~~1~~~~2~~~~2~~~~2~~	
	~~1~~	

Comparing the successive crossing out of the figures in I, II and III, with the rubbing out of figures in the corresponding steps according to the Hindu plan, it becomes quite clear that the galley method is exactly the same as the Hindu method. The crossing out of figures appears to be more cumbrous than the elegant Hindu plan of rubbing out.

The Hindu plan of moving the divisor as the digits of the quotient were evolved, although not essential, was also copied and occurs in the works of such well-known Arab writers as Al-Khowârîzmî (825), Al-Nasavî (*c.* 1025) and others. The mediæval Latin writers called this feature the *antirioratio*.

[1] For details see Smith, *l.c.*, pp. 136-139.

6. SQUARE

Terminology. The Sanskrit term for square is *varga* or *kṛti*. The word *varga* literally means "rows" or "troops" (of similar things). But in mathematics it ordinarily denotes the square power and also the square figure or its area. Thus Āryabhaṭa I says:[1]

"A square figure of four equal sides[2] and the (number representing its) area are called *varga*. The product of two equal quantities is also *varga*."

How the word *varga* came to be used in that sense has been clearly indicated by Thibaut. He says: "The origin of the term is clearly to be sought for in the graphical representation of a square, which was divided in as many *vargas* or troops of small squares, as the side contained units of some measure. So the square drawn with a side of five *padas* length could be divided into five small *vargas* each containing five small squares, the side of which was one *pada* long."[3] This explanation of the origin of the term *varga* is confirmed by certain passages in the *Śulba* works.[4]

The term *kṛti* literally means "doing," "making" or "action." It carries with it the idea of specific performance, probably the graphical representation.

Both the terms *varga* and *kṛti* have been used in the mathematical treatises, but preference is given to the term *varga*. Later writers, while defining these terms in arithmetic, restrict its meaning. Thus Śrīdhara says:[5]

[1] *Ā*, ii. 3.

[2] The commentator Parameśvara remarks: "That four sided figure whose sides are equal and both of whose diagonals are also equal is called *samacaturaśra* ("square")."

[3] Thibaut, *Śulba-sûtras*, p. 48.

[4] *ĀpŚl*, iii. 7; *KŚl*, iii. 9; cf. B. Datta, *American Math. Monthly*, XXXIII, 1931, p. 375.

[5] *Triś*, p. 5.

"The product of two equal numbers is *varga*."

Pṛthudakasvâmî[1], Mahâvîra[2] and others give similar definitions.

The Operation. The occurrence of squaring as an elementary operation is characteristic of Hindu arithmetic. The method, however, is not simpler than direct multiplication. It was given prominence by the Hindu writers probably because the operation of square-root is the exact inverse of that of squaring. Although the method first occurs in the *Brâhma-sphuṭa-siddhânta*, there is no doubt that it was known to Âryabhaṭa I as he has given the square-root method.

Brahmagupta gives the method[3] very concisely thus:

"Combining the product, twice the digit in the less[4] (lowest) place into the several others (digits), with its (*i.e.*, of the digit in the lowest place) square (repeatedly) gives the square."

Śrîdhara (750) is more explicit:[5]

"Having squared the last digit multiply the rest of the digits by twice the last; then move the rest of the digits. Continue the process of moving (the remain-

[1] *Cf.* Colebrooke, *l.c.*, p. 279.

[2] *GSS*, p. 12.

[3] The method is not mentioned in the chapter on Arithmetic, but seems to have been mentioned as an afterthought in the form of an appendix, (*BrSpSi*, p. 212).

[4] *Râśerûnam* has been translated by Colebrooke as "the less portion." This translation is incorrect. He says that "the text is obscure" (p. 322, fn. 9), for according to his translation the rule becomes practically meaningless. The term *râśerûnam* must be translated by "the digit in the lowest place." Dvivedi agrees with the above interpretation (p. 212). The method taught here is "the direct method of squaring."

[5] *Triś*, p. 5. The translation given by Kaye and Ramanujacharia is incorrect. (*Bibl Math* XIII, 1912-13).

ing digits after each operation) to obtain the square."

Mahâvîra[1] (850) gives more details:

"Having squared the last (digit), multiply the rest of the digits by twice the last, (which is) moved forward (by one place). Then moving the remaining digits continue the same operation (process). This gives the square."

Bhâskara II[2] writes:

"Place the square of the last (digit) over itself; and then the products of twice the last (digit) and the others (*i.e.*, the rest) over themselves respectively. Next, moving the number obtained by leaving the last digit (figure), repeat the procedure."

He has remarked that the above process may be begun also with the units place.[3]

The following is the method of working on the *pâṭî*, the process beginning from the last place, according to Śrîdhara, Mahâvîra, Bhâskara II and others:

To square 125.

The number is written down,

125

The last digit is 1. Its square is placed over itself.

1
125

Then twice the last digit $2 \times 1 = 2$; placing it below the rest of the figures (below 2 or below 5 according as the direct or inverse method of multiplication is used)

[1] *GSS*, p. 12.
[2] *L*, p. 4.
[3] *L*, p. 5.

and rubbing out the last digit 1, the work on the *pâṭî* appears as

1
25
2

Performing multiplication by 2 (below) and placing the results over the respective figures, we get

150
25

One round of operation is completed. Next moving the remaining digits, *i.e.*, 25, we have

150
25

Now, the process is repeated, *i.e.*, the square of the last digit (2) is placed over itself giving

154
25

Then, placing twice the last digit (*i.e.*, $2 \times 2 = 4$) below the rest of the digits and then rubbing out 2, we have

154
5
4

Performing multiplication, $4 \times 5 = 20$, and placing it over the corresponding figure 5, (*i.e.*, 0 over 5 and 2 carried to the left), the work on the *pâṭî* appears as

1560
5

Thus a second round of operations is completed.

Then moving 5 we have

1560
5

Squaring 5 we get 25, and placing it over 5 (*i.e.*, 5 over 5 and 2 carried to the left) we have

15625
5

As there are no 'remaining figures' the work ends. 5 being rubbed out, the *pâṭî* has

15625,

the required square.

According to Brahmagupta and also Bhâskara II, the work may begin from the lowest place (*i.e.*, the units place). The following method is indicated by Brahmagupta:

To square 125.

The number is written down

125

The square of the digit in the least place, *i.e.*, $5^2=25$ is set over it thus:

25
125

Then, $2\times5=10$ is placed below the other digits, and five is rubbed out, thus:

25
12
10

Multiplying by 10 the rest of the digits, *i.e.*, 12, and setting the product over them (the digits), we have

1225
12
10

Then rubbing out 10 which is not required and moving the rest of the digits, *i.e.*, 12, we have

1225
12

Thus one round of operations is completed.

Again, as before, setting the square of 2 above it and $2 \times 2 = 4$ below 1, we have

1625
1
4

Multiplying the remaining digit 1 by 4, and setting the product above it, we have

5625
1

Then, moving the remaining digit 1, we obtain

5625
1

Thus the second round of operations is completed.

Next setting the square of 1 above it the process is completed, for there are no remaining figures, and the result stands thus:

15625

Minor Methods of Squaring. The identity

$$(i) \qquad n^2 = (n-a)(n+a) + a^2$$

has been mentioned by all Hindu mathematicians as affording a suitable method of squaring in some cases. For instance,

$$15^2 = 10 \times 20 + 25 = 225.$$

Brahmagupta says:

"The product of the sum and the difference of the number (to be squared) and an assumed number plus the square of the assumed number give the square."[1]

Śrīdhara (750) gives it thus:

"The square is equal to the product of the sum and the difference of the given number and an assumed

[1] *BrSpSi*, p. 212.

quantity plus the square of the assumed quantity."[1] Mahâvîra, Bhâskara II, Nârâyaṇa and others also give this identity.

The formula

(*ii*) $(a+b)^2 = a^2+b^2+2ab$,

or its general form

$(a+b+c+ \ldots .)^2 = a^2+b^2+c^2+ \ldots . +2ab+ \ldots$

has been given as a method of squaring. Thus Mahâvîra[2] says:

"The sum of the squares of the two or more portions[3] of the number together with their products each with the others multiplied by two gives the square."

Bhâskara II[4] gives:

"Twice the product of the two parts plus the square of those parts gives the square."

The formula

(*iii*) $n^2=1+3+5+ \ldots .$ to n terms

has been mentioned by Srîdhara and Mahâvîra.

Srîdhara[5] says:

"(The square of a number) is the sum of as many terms in the series of which one is the first term and two the common difference."

[1] *Triś*, p. 5.

[2] *GSS*, p. 13.

[3] The word *sthâna* has been used in the original. This word has been generally used in the sense of 'notational place.' Following the commentator, we have rendered it by "portion." As a given number, say, 125, can be broken into parts as 50+40+35 or as 100+20+5, and as the rule applies to both, it is immaterial whether the word '*sthâna*' is translated by 'place' or 'portion.' This rule appears to have been given as an explanation of the Hindu method of squaring used with the place-value numerals.

[4] *L*, p. 4.

[5] *Triś*, p. 5.

Mahâvîra uses a similar expression:

"The sum of as many terms of the series of which one is the first term and two the common difference is the square (of the given number)."[1]

Nârâyaṇa[2] has also given the above series. He adds the formula.[3]

$$A^2 = (a+b)^2 = (a-b)^2 + 4ab$$

to find the square of the given number A.

It should be observed that all the above methods of squaring are applied to whole numbers only. The methods of squaring fractions are dealt with in the section on fractions.

7. CUBE

Terminology. The Hindu term for the cube is *ghana*. This term occurs in all the mathematical works. It has been used in the geometrical as well as the arithmetical sense, *i.e.*, to denote the solid cube as well as the continued product of the same number taken three times. Thus Âryabhaṭa I says:

"The continued product of three equals and also the solid having twelve (equal) edges are called *ghana*."[4]

Śrîdhara,[5] Mahâvîra[6] and Bhâskara II[7] each state:

"The continued product of three equal numbers is *ghana*."

[1] *GSS*, p. 12.

[2] *GK*, i. 17-18.

[3] He says: "The square of the difference plus four times the product is the square of the sum."

[4] *Â*, ii. 3.

[5] *Triś*, p. 6.

[6] *GSS*, p. 14.

[7] *L*, p. 5.

Another term for the cube is *bṛnda*, but it is seldom used.

The Operation. A method of cubing applicable to numbers written in the decimal place-value notation, has been in use in India from before the fifth century A.D. Aryabhaṭa I (499) knew the method, although he did not consider it to be as important as the inverse operation of extracting the cube-root which he has explained.

Brahmagupta gives the method in these words:

"Set down the cube of the last;[1] then place at the next place from it, thrice the square of the last multiplied by the succeeding; then place at the next place thrice the square of the succeeding multiplied by the last; and (at the next place) the cube of the succeeding. This gives the cube."

Srîdhara[2] says:

"Set down the cube of the last; then set down at the next place[3] the square of the last multiplied by

[1] The commentator Pṛthudakasvâmî takes 'last' (*antya*) to mean the figure or figures on the extreme left. The number to be cubed is divided into portions of which the cubes are found successively by the application of the process. (See illustration). Thus 'succeeding' would mean here the figure to the right of the 'last.' It does not, however, make any difference if we take 'last' to mean the figure in the units place. The words used by Brahmagupta are '*antya*,' for last and '*uttara*' for succeeding. The rule is:

Sthâpyo'ntya ghano'ntya kṛtistriguṇottarsaṅguṇâ ca tatprathamât ।
Uttarakṛtirantyaguṇâ triguṇâ cottaraghanaśca ghanaḥ ॥

[2] *Triś*, p. 6.

[3] *Sthânâdhikyam*, lit. "increasing one place." This increase is to be made by placing the result one place to the right of the previously noted figures (see illustration). Brahmagupta uses the word *tatprathamât*, i.e., "before that."

thrice the succeeding;[1] then (at the next place) the product of the square of the succeeding and last multiplied by three; and then (at the next place) the cube of the succeeding."

Mahâvîra states:[2]

"The cube of the last, the product of thrice its square and the remaining, the square of the remaining multiplied by thrice the last; placing of these, each one place before the other, constitutes here the process."

Bhâskara II is more explicit:[3]

"Set down the cube of the last; then the square of the last multiplied by three times the succeeding; then the square of the succeeding multiplied by three times the last and then the cube of the succeeding; these placed so that there is difference of a place between one result and the next,[4] and added give the cube. The given number is distributed into portions according to places, one of which is taken for the last and the next as the first and in like manner repeatedly (if there be occasion). Or the same process may be begun from the first place of figures for finding the cube."

The method may be illustrated by the following example:

[1] *Pûrva, âdi*, lit. "preceding". We have rendered them by "succeeding" to be in conformity with the general convention so as to avoid confusion.

[2] *GSS*, p. 15 (47).

It will be observed that the "addition of the cube of the remaining" does not occur in the rule. This has to be understood from the previous stanza which says that the cubes of all the parts are to be added. See the translation of the previous stanza given on pp. 166f.

[3] *L*, p, 5.

[4] *Sthânântaratvena* has been translated by Colebrooke by "according to places." This translation is incorrect and does not give the true significance of the term.

Example. To cube 1234.

The given number has four places, *i.e.*, four portions. First we take the last digit 1 and the succeeding digit 2, *i.e.*, 12, and apply the method of cubing thus :

(i) Cube of the last (1^3)	=	1	
(ii) Thrice the square of the last (3.1^2) multiplied by the succeeding (2) gives ($2.3.1^2$)	=	6	(placing at the next place)
(iii) Thrice the square of the succeeding multiplied by the last gives ($3.2^2.1$)	=	12	(placing at the next place)
(iv) Cube of the succeeding (2^3)	=	8	(placing at the next place)
Thus 12^3 is the sum		1728	

After this we take the next figure 3, *i.e.*, the number 123, and in this consider 12 as the last and 3 as the succeeding. Then the method proceeds thus:

(i) The cube of the last (12^3) as already obtained	=	1728	
(ii) Thrice the square of the last multiplied by the succeeding, *i.e.*, $3.12^2.3$	=	1296	(placing at the next place).
(iii) Thrice the square of the succeeding multiplied by the last, *i.e.*, $3.3^2.12$	=	324	,, ,,
(iv) Cube of the succeeding, *i.e.*, 3^2	=	27	,, ,,
Thus 123^3 is the sum		1860867	

Now the remaining figure 4 is taken, so that the number is 1234, of which 123 is the last and 4 the succeeding. The method proceeds thus:

(i) Cube of the last, *i.e.*, $(123)^3$ as already obtained	=	1860867	
(ii) Thrice the square of the last into the succeeding, *i.e.*, $3.(123)^2.4$	=	181548	(placing at the next place)
(iii) Thrice the square of the succeeding into the last, *i.e.*, $3.4^2.123$	=	5904	" "
(iv) Cube of the last *i.e.*, 4^3	=	64	" "
Thus $(1234)^3$ is the sum		1879080904	

The direct process—that in which the operation begins with the units place—can be similarly performed.

Minor Methods of Cubing. The formula

$$(i) \qquad (a+b)^3 = a^3+3a^2b+3ab^2+b^3$$

and the corresponding result

$$(a+b+c+\ldots.)^3 = a^3+3a^2(b+c+\ldots.)+3a(b+c+\ldots)^2 + (b+c+\ldots.)^3$$

are implied in the Hindu method of cubing given above. Mahâvîra[1] gives the following explanation:

"The squares of the last place[2] and the next[3] are taken, and each (square) is multiplied by the other and by three. The sum of these products and the cubes of both (*lit.* all) the places is the cube; the

[1] *GSS*, p. 15.

[2] *Sthâna*, meaning the number represented by the figure standing in that place.

[3] *Anya*, lit. "other," meaning the number represented by the figures standing in the other places.

procedure is repeated (if necessary).[1]

Śrîpati and Bhâskara II[2] state the formula in the form

$$(a+b)^3 = a^3+3ab\,(a+b)+b^3$$

"Thrice the given number multiplied by its two parts, added to the sum of the cubes of those parts, gives the cube."

Nârâyaṇa[3] says

"Thrice the (given) number multiplied by both parts, added to the cubes of the parts, is the cube of the sum."

The formula

$$(ii) \qquad n^3 = n(n+a)(n-a)+a^2(n-a)+a^3$$

has been mentioned by Mâhâvîra[4] in these words:

"The continued product of the given number, the sum and the difference of the given number and an arbitrary quantity, when added to the smaller of these multiplied by the square of the arbitrary number, and the cube of the arbitrary number, give the cube (of the given number)."

Expressions for n^3 involving series have been given by Srîdhara, Mahâvîra, Śrîpati and Nârâyaṇa. The formula

$$(iii) \qquad n^3 = \sum_{1}^{n} \left\{ 3r\,(r-1)+1 \right\}$$

[1] Thus $(234)^3$ is considered as

$$(200+30+4)^3=(200)^3+3.200^2(30+4)+3.200(30+4)^2 +(30+4)^3$$

Then the procedure is repeated for obtaining $(30+4)^3$. *Cf.* English translation, p. 17, note.

[2] *GT*, 27; *L*, p. 5.

[3] *GK*, i. 23.

[4] *GSS*, p. 15.

is given by Srîdhara in these words:

"The cube (of a given number) is equal to the series whose terms are formed by applying the rule, 'the last term multiplied by thrice the preceding term plus one,' to the terms of the series whose first term is zero, the common difference is one and the last term is the given number."[1]

Mahâvîra gives the above in the form

$$n^3 = 3\sum_{2}^{n} r(r-1)+n$$

He[2] says:

"In the series, wherein one is the first term as well as the common difference and the number of terms is equal to the given number (n), multiply the preceding term by the immediately following one. The sum of the products so obtained, when multiplied by three and added to the last term (*i.e.*, n) becomes the cube (of n)."

Nârâyaṇa[3] states:

"From the series whose first term and common difference are each one, (the last term being the given number) the sum of the series formed by the last term multiplied by three and the preceding added to one, gives the cube (of the last term)."

Mahâvîra has also mentioned the results,

(*iv*) $x^3 = x+3x+5x+ \ldots$ to x terms,

(*v*) $x^3 = x^2+(x-1)\{1+3+\ldots+(2x-1)\}$,

[1] *Triś*, p. 6. The translation given by Kaye and Ramanujacharia (*Bibl.Math.*, III, 1912-13) is incorrect. They admit their inability to follow the meaning (see p. 209, note). S. Dvivedi has misinterpreted the rule, and gives an incorrect explanation in a note on p. 6. The reading *saike* is incorrect.

[2] *GSS*, ii. 45.

[3] *GK*, i. 22.

in these words:[1]

"The cube (of a given number) is equal to the sum of the series whose first term is the given number, the common difference is twice that number, and the number of terms is (equal to) that number.

"Or the square of the given number when added to the product of that number minus one (and) the sum of the series in which the first term is one, the common difference two and the number of terms (is equal to) that number, gives the cube."

8. SQUARE-ROOT

Terminology. The Hindu terms for the "root" are *mûla* and *pada*. The usual meaning of the word *mûla* in Sanskrit literature is "root" of a plant or tree; but figuratively the foot or lowest part or bottom of anything. Its other meanings are "basis," "foundation," "cause," "origin," etc. The word *pada* means "the lower part of the leg" (figuratively the lower part or basis of anything), "foot," "part," "portion," "side," "place," "cause," "a square on a chess-board," etc. The meanings common to both terms are "foot," "the lowest part or basis of anything," "cause" or "origin." It is, therefore, quite clear that the Hindus meant by the term *varga-mûla* ("square-root") "the cause or origin of the square" or "the side of the square (figure)." This is corroborated by the following statement of Brahmagupta:[2]

"The *pada* (root) of a *kṛti* (square) is that of which it is the square."

Of the above terms for the "root," *mûla* is the oldest. It occurs in the *Anuyogadvâra-sûtra* (*c*. 100 B.C.),

[1] *GSS*, ii. 44.
[2] *BrSpSi*, xviii. 35.

and in all the mathematical works. The term *pada* seems to have come into use later on, *i.e.*, from the seventh century A.D. It occurs first in the work of Brahmagupta (628).

The term *mûla* was borrowed by the Arabs who translated it by *jadhr*, meaning "basis of square." The Latin term *radix* also is a translation of the term *mûla*[1].

The word *karanî* is found to have been used in the *Śulba* works and Prâkrta literature as a term for the square-root. In geometry it means a "side." In later times the term is, however, reserved for a surd, *i.e.*, a square-root which cannot be evaluated, but which may be represented by a line.

The Operation. The description of the method of finding the square-root is given in the *Âryabhaṭîya* very concisely thus:

"Always divide[2] the even place by twice the square-root (upto the preceding odd place); after having subtracted from the odd place the square[3] (of the quotient), the quotient put down at the next place (in the line of the root) gives the root."[4]

The method may be illustrated thus:

Example. Find the square-root of 54756.

[1] For further details see Datta, *American Math. Monthly*, XXXIV, pp. 420-423, also XXXVIII, pp. 371-376.

[2] In dividing, the quotient should be taken as great as will allow of the subtraction of its square from the next odd place. This is the force of the Sanskrit text as pointed out by the commentators Bhâskara I, Nîlakantha and others.

[3] The "square" is mentioned and not the "square of the quotient," as in the beginning the greatest possible square is to be subtracted, there being no quotient.

[4] *Â*, ii. 4. Translations of the rule have been given before by Rodet (*JA*, 1880, II), Kaye (*JASB*, 1907 and 1908, III and IV resp.), Singh (*BCMS*, 1927, XVIII), Clark (*Âryabhaṭîya*) and others. Of these Kaye's translation is entirely incorrect.

The odd and even places are marked by vertical and horizontal lines. The different steps are then as indicated below:

	| ‾ | ‾ |	
	5 4 7 5 6	
Subract square	4	root=2
Divide by twice the root	4) 14 (3	Placing quotient at the
	12	next place, the root =23
Subtract square of quotient	27	
	9	
Divide by twice the root	46) 185 (4	Placing quotient at the
	184	next place, the root =234
Subtract square of quotient	16	
	16	

The process ends. The root is 234.

It has been stated by G. R. Kaye[1] that Âryabhaṭa's method is algebraic in character, and that it resembles the method given by Theon of Alexandria. Both his statements are incorrect.[2]

The following quotations from Siddhasena Gaṇi (*c.* 550) in his commentary on the *Tatvârthâdhigama-sûtra*[3] will prove conclusively that the Hindu method of extracting the square-root was not algebraic. In connection with the determination of the circumference of a circle of 100,000 *yojanas*, he says:

"The diameter is one hundred thousand *yojanas*; that multiplied by one hundred thousand *yojanas* becomes squared; this is again multiplied by 10 and then

[1] *JASB*, III and IV, in the papers entitled "Notes on Indian Mathematics, I and II."

[2] See Singh, *l.c.*, for details; also Clark, *Âryabhaṭîya*, pp. 23f.

[3] iii. 11.

the square-root (of the product) extracted. The root will be the circumference of the circle. Now to find the number of *yojanas* (by extracting the square-root) we obtain in succession the figures 3,1,6,2,2 and 7 of the root, the number appearing below (that is, as the last divisor) is 632454. This being halved becomes the number three hundred thousand sixteen thousands two hundred and twenty seven. The number in excess as the remainder is this 484471;...."

"Then on multiplication by 4 will be obtained 75600000000. The square-root of this will be the chord. In finding that (root) will be obtained in succession the figures 2,7,4,9,5 and 4;...."

It is evident that Âryabhaṭa's plan of finding the square-root has been followed in the above cases as the digits of the root are evolved successively one by one.

Later writers give more details of the process. Thus Śrîdhara says:

"Having subtracted the square from the odd place, divide the next (even) place by twice the root which has been separately placed (in a line), and after having subtracted the square of the quotient, write it down in the line; double what has been obtained above (by placing the quotient in the line) and taking this down, divide by it the next even place. Halve the doubled quantity (to get the root)."[1]

Mahâvîra,[2] Âryabhaṭa II[3] and Śrîpati[4] give the rule in the same way as Śrîdhara. Bhâskara II, however, makes a slight variation, for he says:

[1] *Triś*, p. 5. For an illustration of the method of working on a *pâṭî*, see A. N. Singh, *BCMS*, XVIII, p. 129.

[2] *GSS*, p. 13.

[3] *MSi*, p. 145.

[4] *SiŚe*, xiii. 5; *GT*, 23.

"Subtract from the last odd place the greatest square number. Set down double the root in a line, and after dividing by it the next even place subtract the square of the quotient from the next odd place and set down double the quotient in the line. Thus repeat the operation throughout all the figures. Half of the number in the line is the root."[1]

The method of working on the *pâṭî* may be illustrated as below:

Example. Find the square-root of 54756

The given number is written down on the *pâṭî* and the odd and even places are marked by vertical and horizontal lines thus:

$$\overset{|}{5}\,\overset{-}{4}\,\overset{|}{7}\,\overset{-}{5}\,\overset{|}{6}$$

Beginning with the last odd place 5, the greatest square number 4 is subtracted. Thus 4 subtracted from 5 gives 1. The number 5 is rubbed out and the remainder 1 substituted in its place. Thus after the first operation is performed, what stands on the *pâṭî* is

$$1\,\overset{-}{4}\,\overset{|}{7}\,\overset{-}{5}\,\overset{|}{6}$$

Double the root 2, *i.e.*, 4, is permanently placed in a separate portion of the *pâṭî* which has been termed *paṅkti* ("line"). Dividing the number upto the next even mark by this number in the line, *i.e.*, dividing 14 by 4 we obtain the quotient 3 and remainder 2. The number 14 is rubbed out and the remainder 2 written

[1] *L*, p. 4. The line in Bhâskara II's method contains the doubled root, whilst in that of Âryabhaṭa I, it contains the root. See Singh, *l.c.*

in its place; thus on the *pâṭî* we have now

4	27 56	(3 quotient
line of root		

The square of the quotient $3^2=9$ is subtracted from the figures upto the next odd mark. This gives $(27-9)=18$. 27 is rubbed out and 18 substituted in its place. Double the quotient 3 is now set in the line giving 46. The figures on the *pâṭî* stand thus:

46	1856	The quotient 3 having been rubbed out.
line of root		

Dividing the numbers upto the next even mark by the number in the line, *i.e.*, dividing 185 by 46, the quotient is 4 and remainder 1. 185 is rubbed out and the remainder 1 substituted in its place. The figures on the *pâṭî* are now

46	16	(4 quotient
line of root		

Subtracting square of the quotient the remainder is nil, so that 16 is rubbed out. The quotient 4 is doubled and set in the line. The *pâṭî* has now

468	The quotient 4 having been rubbed out.
line of root	

Half the number in the line, *i.e.*, $\frac{468}{2}=234$ is the root.

Along with the Hindu numerals, the method of extracting the square-root given above, seems to have been communicated to the Arabs about the middle of the eighth century, for it occurs in precisely the same form in Arabic works on mathematics.[1] In Europe

[1] *E.g.*, Al-Nasavî (1025); see Suter, *Bibl. Math.*, VII, p. 114 and Woepcke, *JA* (6), t. 1, 1863.

it occurs in similar form in the writings of Peurbach (1423-1461), Chuquet (1484), La Roche (1520), Gemma Frisius (1540), Cataneo (1546) and others.[1]

9. CUBE-ROOT

Terminology. The Hindu terms for the cube-root are *ghana-mûla*, *ghana-pada*. These terms have already been discussed before.

The Operation. The first description of the operation of the cube-root is found in the *Aryabhaṭîya*. It is rather too concise:

"Divide the second *aghana* place by thrice the square of the cube-root; subtract from the first *aghana* place the square of the quotient multiplied by thrice the preceding (cube-root); and (subtract) the cube (of the quotient) from the *ghana* place; (the quotient put down at the next place (in the line of the root) gives the root)."[2]

As has been explained by all the commentators, the units place is *ghana*, the tens place is first *aghana*, the hundreds place is second *aghana*, the thousands place is *ghana*, the ten-thousands place is first *aghana* and so on. After marking the places as *ghana*, first *aghana* and second *aghana*, the process begins with the subtraction of the greatest cube number from the figures upto the last *ghana* place. Though this has not been ex-

[1] See Smith, *History*, II, pp. 144-148.

[2] *A*, ii. 5. Translations of this rule have been given by Rodet, Kaye, Singh, Clark, Sengupta and others. Kaye's translation is entirely inaccurate. Other translations, though free, give the correct result. Clark's use of the words "the (preceding) *ghana*" is somewhat misleading. The portion at the end, within brackets, is common to this and the preceding rule for the extraction of the square-root.

plicitly mentioned in the rule, the commentators say that it is implied in the expression "*ghanasya mûla vargeṇa*" etc. ("by the square of the cube-root" etc.) The method may be illustrated as below:

Example. Find the cube-root of 1953125.

The places are divided into groups of three by marking them as below:

	|--|--| 1953125		
Subtract cube	1	(c)	Root = 1
Divide by thrice square of root, *i.e.* 3.1^2	3) 9 (2....	(a)[1]	Placing quotient after the root 1 gives the root 12
Subtract square of quotient multiplied by thrice the previous root, *i. e.*, $2^2.\ 3.\ 1.$	6 —— 35 12	(b)	
Subtract cube of quotient, *i.e.*, 2^3	—— 233 8	(c)	
Divide by thrice square of the root, *i.e.*, $3.\ 12^2$	—— 432) 2251 (5 .	(a)	Placing quotient after the root 12 gives the root 125
Subtract square of quotient multiplied by thrice the previous root, *i.e.*, $5^2.\ 3.\ 12$	2160 —— 912 900 ..	(b)	
Subtract cube of quotient, *i.e.*, 5^3	—— 125 125 ..	(c)	
	——		

Thus the cube-root = 125

It will be evident from the above illustration that the present method of extracting the cube-root is a contraction of Āryabhaṭa's method.

The method given above occurs in all the Hindu mathematical works. For instance, Brahmagupta says:

[1]The quotient in division is to be taken as great as will allow the two subsequent operations (*b*) and (*c*) to be carried out.

"The divisor for the second *aghana* place is thrice the square of the cube-root; the square of the quotient multiplied by three and the preceding (root) must be subtracted from the next (*aghana* place to the right), and the cube (of the quotient) from the *ghana* place; (the procedure repeated gives) the root."[1]

Srîdhara gives more details of the process as actually performed on the *pâṭî*, thus:

"(Divide the digits beginning with the units place into periods of) one *ghana* place and two *aghana* places. From the (last) *ghana* digit subtract the (greatest possible) cube; then taking down the remainder and the third *pada* (*i.e.*, the second *aghana* digit) divide it by thrice the square of the cube-root which has been permanently placed in a separate place; place the quotient in the line; subtract the square of this (quotient) multiplied by thrice the last root from the next (*aghana*) digit. Then as before subtract the cube (of the quotient) from its own place (*i.e.*, the *ghana* digit). Then take down again the third *pada* (*i.e.*, second *aghana* digit), and the rest of the process is as before. (This will give) the root."[2]

Aryabhaṭa II follows Srîdhara and gives details as follows:

"*Ghana* (*i.e.*, the place from which cube is subtracted), *bhâjya* (*i.e.*, the "dividend" place) and *śodhya* (*i.e.*, the "minuend" place) are the denominations (of the places). Subtract the (greatest) cube from its own place (*i.e.*, from the numbers upto the last *ghana* digit); bring down the *bhâjya* digit and divide it[3] by thrice the square of the cube-root which has been permanently

[1] *BrSpSi*, p. 175; *cf.* Colebrooke, *l.c.*, p. 280.
[2] *Triś*, pp. 6f.
[3] Literally, its own place.

placed. Place the quotient in the line (of the root). The square of this (quotient) multiplied by thrice the previous root is subtracted from its own place (*i.e.*, the *śodhya* place) and its cube from the *ghana* place. If the above operations be possible then this (*i.e.*, the number in the line) is the root so far. Then bringing down the *bhâjya* digit continue the process as before (till it ends)."[1]

The component digits of the number whose cube-root is to be found are divided into groups of three (one *ghana* and two *aghana*s) each. The digits upto the last *ghana* place (proceeding from left to right) give the first figure of the root (counting from the left). The following period of three digits (to the right) gives the second figure of the root and so on. While working on the *pâṭî*, the digits of the number whose root is to be found are first marked and the method proceeds as follows:

Example. Find the cube-root of 1953125.

The number is written thus:

```
|--|--|
1953125
```

From the last *ghana* digit (marked by a vertical stroke), the greatest cube is subtracted. Here 1^3 being subtracted from 1 gives zero. So 1 is rubbed out. The cube-root of 1^3 is placed in a separate line. The figures on the *pâṭî* stand thus:

```
--|--|          1
953125    ------------
          line of root
```

Then to obtain the second figure of the root, the second *aghana* (*i.e.*, 9) is taken below and divided by

[1] *MSi*, p. 145. The interpretation given by Dvivedi of line 2 of the rule as printed in his edition (p. 145) is incorrect.

thrice the square of the root (*i.e.*, the number in the line). Thus we have

```
                --|--|
                953125
3.1²    =    3) 9     (2    quotient
                6
                -
                3
```

The quotient is taken to be 2, because if it were taken to be 3, the rest of the procedure cannot be carried out. This quotient (2) is set in the line. The first *aghana* is then brought down and we have, on subtracting the square of the quotient multiplied by thrice the previous root, the following:

```
                --|--|                       12
                953125                   ------------
3.1² =       3) 9     (2   quotient      line of root
                6
                --
                35
2². 3.1 =       12
                --
                23
```

On bringing down the *ghana* digit 3, and then subtracting the cube of the quotient we get 225 as below, and the process on the period formed by the digits 953 is completed and the figure 2 of the root is obtained:

```
                --|--|
                953125
             3) 9     (2              12
                6                 ------------
                --                line of root
                35
                12
                ---
                233
2³ =              8
                ---
                225
```

The figures 953 are then rubbed out and the remainder 225 is substituted. After this the process is as before, *i.e.*, thus

	- - |	
	225125	
$12^2 . 3 =$	432) 2251 (5	12
	2160	line of root
	912	
$5^2 . 12 . 3 =$	900	
	125	
$5^3 =$	125	125
	0	line of root

The process ends as all the figures in the number are exhausted. The root is 125, the number in the line of root. As there is no remainder, the root is exact.

The necessity for rubbing out figures arises, as the *pâṭî* is not big enough to contain the whole of the working. The three digits constituting a period are considered together. The figures upto the next second *aghana* have to be brought down and the operation of division performed separately, because the quotient is obtained by trial. As has been already explained, this division is performed by rubbing out the digits of the dividend (and not as in the working given above). If the operations were carried out on the figures of the original number, and if the quotient taken were found to be too big, then it would not be possible to restore the original figures and begin the work again, as will have to be done in case of failure.

10. CHECKS ON OPERATIONS

The earliest available description of a method of checking the results of arithmetical operations, the direct as well as the inverse, is found in the *Mahâsiddhânta*[1] (*c.* 950). It says:

"Add together the own digits of the numbers forming the multiplicand, multiplier, and product upto one place;[2] such should be done with the dividend, divisor, quotient and remainder, etc. Then if the number (of one digit) obtained from the product of those numbers (that have been already obtained) from the multiplicand and the multiplier be equal to that obtained from the product, the multiplication is true. If the number, which results from the product of those obtained from the quotient and the divisor, added to that from the remainder, be equal to that obtained from the dividend, the division is true. Add together the digits of a number, its (nearest) square-root (in integers) and of the remainder. If the number, obtained from the square of that (number) which is obtained from the square-root plus the number obtained from the remainder, be equal to that which results from the given number, the root-extraction is true. If the number, resulting from the cube of the number obtained by adding the digits of the cube-root plus the number obtained from the remainder, be equal to the number resulting from the given number, then the operation is correct. Such are the easy tests for correctness of multiplication etc."

The *rationale* of the above rules will be clear from the following: Let

$$n = d_m d_{m-1} \ldots\ldots d_2 d_1$$

be a number of m digits written in the decimal place-value notation. Let S_1 denote the sum of its digits,

[1] *MSi*, p. 452.

[2] That is, the digits of the number should be added together; the digits of the sum thus obtained should be again added and the process should be continued until there remains a number of one digit only.

S_2 the sum of the digits of S_1, and so on.

Then

$$n = d_1 + 10d_2 + \ldots\ldots + 10^{m-1}d_m,$$
$$S_1 = d_1 + d_2 + d_3 + \ldots\ldots + d_m$$

so that

$$n - S_1 = 9(d_2 + 11d_3 + \ldots\ldots).$$

Therefore,

$$n \equiv S_1 \pmod{9},$$

Similarly

$$S_1 \equiv S_2 \pmod{9},$$
$$S_2 \equiv S_3 \pmod{9},$$
$$\ldots\ldots\ldots\ldots$$
$$\ldots\ldots\ldots\ldots$$
$$S_{k-1} \equiv S_k \pmod{9},$$

where S_k is a number of one digit only, say n', which is certainly less than or equal to 9.

Adding the congruences, we obtain

$$n \equiv n' \pmod{9}.$$

Thus the number obtained by adding the digits of a number repeatedly is equal to the remainder obtained by dividing that number by nine.

Now, if there be a number N which is equal to the continued product of p other numbers $n_1, n_2, n_3, \ldots, n_p$ plus or minus another number R, then we write

$$N = n_1.n_2.n_3 \ldots n_p \pm R$$

Now, let

$$n_1 \equiv n'_1 \pmod{9}$$
$$n_2 \equiv n'_2 \pmod{9}$$
$$\ldots\ldots\ldots\ldots$$
$$\ldots\ldots\ldots\ldots$$
$$n_p \equiv n'_p \pmod{9}$$

Multiplying the congruences, we obtain

$$n.n_2 \ldots . n_p \equiv n'_1.n'_2 \ldots . n'_p \text{ (mod. 9).}$$

Further let

$$R \equiv r' \qquad \text{(mod. 9)}$$

Therefore

$$n_1.n_2.n_3 \ldots . n_p \pm R \equiv n'_1.n'_2 \ldots . n'_p \pm r' \text{ (mod. 9).}$$

Hence

$$N \equiv n'_1.n'_2 \ldots\ldots n'_p \pm r' \text{ (mod. 9).}$$

In particular, if

$$n_1 = n_2 = \ldots\ldots = n_p = n, \text{ say}$$

then will

$$n'_1 = n'_2 = \ldots\ldots\ldots = n'_p = n'.$$

Therefore,

$$N = n^p \pm R$$

and

$$N \equiv n'^p \pm r' \text{ (mod. 9).}$$

From the above follow easily the rules of the *Mahâsiddhânta.*

The following rule for testing multiplication is given by Nârâyaṇa[1] (1356):

"The remainders obtained on division of each of the multiplicand and the multiplier by an optional number are multiplied together and then divided by the optional number. If the remainder so obtained be equal to the remainder obtained on dividing the product (of the multiplicand and the multiplier) by the optional number, then, it is correct."

It must be noted here that a complete set of rules for checking by nines is first found in India. Methods for testing multiplication and division were probably

[1] Quoted by S. Dvivedi, *History of Mathematics* (in Hindi), Benares, 1910, p. 79.

known to the Hindus much earlier. But as these tests were not considered to be among the fundamental operations, they were not mentioned in the works on *pâṭî-gaṇita*.[1] Nârâyaṇa seems to be the first Hindu mathematician to give rules for testing operations by the casting out of any desired number.

In the works of early Arab writers the methods of testing multiplication, and division without remainder, by the check of nines are given, while a complete set of rules for testing all operations is found first in the works of Avicenna[2] (*c.* 1020) who calls his method the "Hindu" method. Maximus Planudes[3] also ascribes an Indian origin to the check of nines.

There is thus no doubt as to the Hindu origin of the check of nines. Before Âryabhaṭa II, it was probably used to test multiplication and division only. It was perhaps in this imperfect form when it was communicated to the Arabs. Thereafter, the method was probably perfected independently both in Arabia and India. This would account for the difference in the formulation of the rules by the Arabs and by Âryabhaṭa II, the author of the *Mahâsiddhânta*.[4] It is, however, certain that the Hindus did not borrow the method from the Arabs, because Aryabhaṭa II wrote before Avicenna. Behâ Eddîn[5] (*c.* 1600) gives the check of nines in exactly the same form as Aryabhaṭa II.

[1] Besides the above works, the check of nines is also quoted by Sumatiharṣa (1618) from an anterior writer Bîjadatta, in his commentary on the *Karaṇa-kutûhala* of Bhâskara II, ed. by Mâdhava Śâstrî, Bombay, 1901, i. 7.

[2] F. Woepcke, *JA*(6), I, 1863, pp. 500 *et sq.*

[3] *Vide* Delambre, *Histoire de l'Astronomie Ancienne*, t. I, Paris, 1817, pp. 518 ff.

[4] Noted by B. Datta, *JASB*, XXIII, 1927, p. 265.

[5] *Kholâsat al-hisâb*, French translation by A. Marre, *Nouvelles Annales d. Math.*, t. v, 1846, p. 263.

11. FRACTIONS

Early Use. In India, the knowledge of fractions can be traced back to very early times. In the oldest known work, the *Ṛgveda,* the fractions one-half (*ardha*) and three-fourths (*tri-pâda*[1]) occur. In a passage of the *Maitrâyaṇî Saṁhitâ*[2] are mentioned the fractions one-sixteenth (*kalâ*), one-twelfth (*kuṣṭha*), one-eighth (*śapha*) and one-fourth (*pâda*). In the earliest known mathematical works, the *Sulba-sûtra*, fractions have not only been mentioned, but have been used in the statement and solution of problems.[3]

The ancient Egyptians and Babylonians are known to have used fractions with unit numerators, but there is little evidence of the use by these people of what are called composite fractions. The occurrence of the fraction three-fourths in the *Ṛgveda* is probably the oldest record of a composite fraction known to us. The Sanskrit compound *tri-pâda* literally means "three-feet." Used as a number it denotes that the measure of the part considered bears the same ratio to the whole as three feet of a quadruped bear to the total number of its feet. The term *pâda*, however, is a word numeral for one-fourth, and the compound *tri-pâda* is formed exactly on the same principle as the English term three-fourths.[4] In the *Sulba*, unit fractions are denoted by the use of a cardinal number with the term *bhâga* or *aṁśa*; thus *pañca-daśa-bhâga* ("fifteen-parts") is equivalent to one-fifteenth,[5] *sapta-bhâga* ("seven-parts") is equivalent to one-seventh,[6] and so on. The use of ordinal numbers

[1] *RV*, x. 90. 4.
[2] iii. 7. 7.
[3] B. Datta, *Śulba*, pp. 212 ff.
[4] *tri*=three and *pâda*=fourth.
[5] *ÂpŚl*, x. 3; *KŚl*, v. 8.
[6] *KŚl*, vi. 4.

with the term *bhâga* or *aṁśa* is also quite common, e.g., *pañcama-bhâga* ("fifth part") is equivalent to one-fifth.[1] Sometimes the word *bhâga* is omitted, probably for the sake of metrical convenience.[2] Composite fractions like 3/8 and 2/7 are called *tri-aṣṭama* ("three-eighths") and *dvi-saptama* ("two-sevenths") respectively. In the Bakhshâlî Manuscript the fraction 3/8 is called *tryaṣṭa* ("three-eighths") and $3\frac{3}{8}$ is called *trayastrayaṣṭa* ("three-three-eighths").[3] Instances of the formation of fraction names on the above principle are too numerous in later works to be mentioned here. The present method of expressing fractions is thus derived from Hindu sources and can be traced back to 3,000 B.C.

Weights and Measures. The division of the units of length, weight, money, etc., into smaller units for the sake of avoiding the use of fractional quantities has been common amongst all civilised peoples. It is an index of commercial activity and the development of commercial arithmetic. The Hindus have used systems of weights and measures from the earliest times. The *Śatapatha Brâhmaṇa*[4] (*c.* 2,000 B.C.) gives a very mintute subdivision of time. According to it there are 30 *muhûrta* in a day, 15 *kṣipra* in a *muhûrta*, 15 *itarhi* in a *kṣipra*, 15 *idânî* in an *itarhi* and 15 *prâṇa* in an *idânî*. Thus one *prâṇa* is approximately equivalent to one-seventeenth of a second. It is improbable that the ancient Hindus had any appliance for measuring such small intervals of time. The subdivision is entirely theoretical, and probably made for philosophical reasons. It, nevertheless, shows that the Hindus must

[1] *ÂpŚl*, ix. 7, x. 2; *KŚl*, v. 6.

[2] When the fractions have unit numerators, only the denominators are mentioned. This practice is quite common in later works also, e.g., *ṣaṣṭa* (sixth)=$\frac{1}{6}$ in *L*, p. 7 etc.

[3] *BMs*, 10 *verso*.

[4] xii, 3. 2. 1.

have been in possession of a satisfactory arithmetic of fractions even in those early times. The *Arthaśâstra* of Kauṭilya[1] contains a section dealing with weights and measures which were in use in India in the fourth century B.C. In the *Lâlitavistara*[2] Buddha is stated to have given the following system of linear measures:

7	*paramâṇu raja*	= 1	*reṇu*
7	*reṇu*	= 1	*truṭi*
7	*truṭi*	= 1	*vâtâyana raja*
7	*vâtâyana raja*	= 1	*śaśa raja*
7	*śaśa raja*	= 1	*eḍaka raja*
7	*eḍaka raja*	= 1	*go raja*
7	*go raja*	= 1	*likṣâ raja*
7	*likṣâ raja*	= 1	*sarṣapa*
7	*sarṣapa*	= 1	*yava* (breadth of barley)
7	*yava*	= 1	*aṅgulî parva* (breadth of finger)
12	*aṅgulî parva*	= 1	*vitasti*
2	*vitasti*	= 1	*hasta* (cubit)
4	*hasta*	= 1	*dhanu*
1000	*dhanu*	= 1	*krośa*
4	*krośa*	= 1	*yojana*

According to the above table, the smallest Hindu measure of length, a *paramâṇu*[3] $= 1 \cdot 3 \times 7^{-10}$ inches.

All the works on *pâṭîgaṇita* begin with definitions of the weights and measures employed in them. The earlier ones contain a special rule for the reduction of a chain of measures into a proper fraction.[4] It may be mentioned that the systems of weights and measures

[1] The *Arthaśâstra of Kauṭilya*, ed. by R. Shamsastri, Bangalore, 1919.

[2] *Lalitavistara*, ed. R. Mitra, Calcutta, 1877, p. 168.

[3] *Paramâṇu* is the smallest particle of matter. Thus according to the Hindus, the diameter of a molecule is $1 \cdot 3 \times 7^{-10}$.

[4] The process is called *vallî-savarṇana* and occurs in the *Triśatikâ* (p. 12) and the *Gaṇita-tilaka* (p. 39) and not in later works.

given in different works are different from each other. They are the ones current at the time and in the locality in which the book was composed.

Terminology. The Sanskrit term for a fraction is *bhinna*. It means "broken." The European terms *fractio*, *fraction*, *roupt*, *rotto*, or *rocto* etc., are translations of the term *bhinna*, having been derived from the Latin *fractus* (*frangere*) or *ruptus* meaning "broken." The Hindu term *bhinna*, however, had a more general meaning in so far as it included numbers of the form,

$$\left(\frac{a}{b} \pm \frac{c}{d}\right), \ \left(\frac{a}{b} \text{ of } \frac{c}{d}\right), \ \left(\frac{a}{b} \pm \frac{c}{d} \text{ of } \frac{a}{b}\right) \text{ or } \left(a \pm \frac{\dot{b}}{c}\right).$$

These forms were termed *jâti*, *i.e.*, "classes," and the Hindu treatises contain special rules for their reduction to proper fractions. Srîdhara and Mahâvîra each enumerate six *jâti*s, while Brahmagupta gives only five and Bhâskara II following Skandasena reduces the number to four. The need for the division of fractions into classes arose out of the lack of proper symbolism to indicate mathematical operations. The only operational symbol used by the Hindus was a dot[1] for the negative sign.

The other terms employed for the fraction are *bhâga* and *aṁśa*, meaning "part" or "portion." The term *kalâ* which originally, in Vedic times, denoted one-sixteenth came to be later on employed for a fraction. Its earliest use as a term for fraction occurs in the *Sulba* works.

Writing of Fractions. From very early times (*c.* 200 A.D.) the Hindus wrote fractions just as we do now, but without the dividing line. When several fractions occurred in the same problem, they were in general separated from each other by vertical and

[1] Generally placed over the number to be subtracted.

horizontal lines. Illustrations of the Hindu method of writing groups of fractions will be found in the examples that will be given hereafter.

Reduction to Lowest Terms. Before performing operations with fractions, it was considered necessary to reduce them to lowest terms. The process of reduction was called *apavartana*, but was not included among the operations. It is not given in the Hindu works, but seems to have been taught by oral instruction. That the method has been in use in India from the earliest centuries of the Christian era, cannot be doubted; for it is mentioned in a non-mathematical work, the *Tattvârthâdhigama-sûtra-bhâṣya*[1] by Umâsvâti (*c.* 150) as a simile to illustrate a philosophical discussion:

"Or, as when the expert mathematician, for the purpose of simplifying operations, removes common factors from the numerator and denominator of a fraction, there is no change in the value of the fraction, so . . . . . "

Reduction to Common Denominator. The operation of reduction to a common denominator[2] is required when fractions are to be added or subtracted. The process is given a prominent place and is generally mentioned along with addition and subtraction. Bṛhmagupta[3] gives the reduction along with the processes of addition and subtraction thus:

"By the multiplication of the numerator and denominator of each of the (fractional) quantities by the other denominators, the quantities are reduced to a common denominator. In addition, the numerators are united. In subtraction their difference is taken."

[1] ii. 52.
[2] *Kalâ-savarṇana* or *savarṇana*, or *samachheda-vidhi*.
[3] *BrSpSi*, p. 172.

Srîdhara[1] says:

"To reduce to a common denominator, multiply the numerator and denominator of each (fraction) by the other denominators."

All other works also contain this rule.

Fractions in Combination. It has already been remarked that due to the lack of proper symbolism, the Hindu mathematicians divide combinations of fractions into four classes. They are:

(1)[2] *Bhâga*, *i.e.*, the form $\left(\frac{a}{b} \pm \frac{c}{d} \pm \frac{e}{f} \pm \cdots\right)$.
usually written as

$$\begin{array}{|c|c|c|} \hline a & c & e \\ b & d & f \\ \hline \end{array} \quad \text{or} \quad \begin{array}{|c|c|c|} \hline a & \cdot c & \cdot e \\ b & d & f \\ \hline \end{array}$$

where the dots denote subtraction.

(2)[3] *Prabhâga*, *i.e.*, the form $\left(\frac{a}{b} \text{ of } \frac{c}{d} \text{ of } \frac{e}{f} \cdots\cdots\right)$,
written as

$$\begin{array}{|c|c|c|} \hline a & c & e \\ b & d & f \\ \hline \end{array}$$

(3)[4] *Bhâgânubandha*, *i.e.*, the form

(*i*) $\left(a + \frac{b}{c}\right)$

or (*ii*) $\frac{p}{q} + \frac{r}{s} \text{ of } \frac{p}{q} + \frac{t}{u} \text{ of } \left(\frac{p}{q} + \frac{r}{s} \text{ of } \frac{p}{q}\right) + \cdots$

[1] *Triś*, p. 10. The translation given by Kaye is incorrect.

[2] *BrSpSi*, p. 175; *Triś*, p. 10; *GSS*, p. 33 (55, 56); *MSi*, p. 146; *L*, p. 6.

[3] *Triś*, p. 10; *GSS*, p. 39 (99); *MSi*, p. 146; *L*, p. 6.

[4] *Triś*, p. 10; *GSS*, p. 41 (113); *MSi*, p. 148; *L*, p. 7. These forms are termed *rûpa-bhâgânubandha* ("association of an integer and a fraction") and *bhâga-bhâgânubandha* ("association of fractions of fractions") respectively.

written as

$$(i)\ \begin{array}{|c|}\hline a \\ b \\ c \\ \hline\end{array} \quad \text{or} \quad (ii)\ \begin{array}{|c|}\hline p \\ q \\ \hline r \\ s \\ \hline t \\ u \\ \hline\end{array}$$

(4)[1] *Bhâgâpavâha*, *i.e.*, the form

$$(i)\ \left(a - \frac{b}{c}\right)$$

$$\text{or}\ (ii)\ \frac{p}{q} - \frac{r}{s} \text{ of } \frac{p}{q} - \frac{t}{u} \text{ of } \left(\frac{p}{q} - \frac{r}{s} \text{ of } \frac{p}{q}\right) - \ldots.$$

written as

$$(i)\ \begin{array}{|c|}\hline a \\ \cdot b \\ c \\ \hline\end{array} \quad \text{or} \quad (ii)\ \begin{array}{|c|}\hline p \\ q \\ \hline \cdot r \\ s \\ \hline \cdot t \\ u \\ \hline\end{array}$$

Besides the above four forms, Śrîdhara, Mahâvîra, and some others give two more.

(5)[2] *Bhâga-bhâga*, *i.e.*, the form

$$\left(a \div \frac{b}{c}\right) \text{ or } \left(\frac{p}{q} \div \frac{r}{s}\right)$$

There does not appear to have been any notation for division, such compounds being written as,

[1] *BrSpSi*, p. 176; *GSS*, p. 43 (126); *MSi*, p. 148; *L*, p. 7. These forms are termed *rûpa-bhâgâpavâha* and *bhâga-bhâgâpavâha* respectively.

[2] *Triś*, p. 11; *GSS*, p. 39 (99).

a		*p*
b	or	*q*
c		*r*
		s

just as for *bhâgânubandha*. That division is to be performed was known from the problem;[1] *e.g.*, $1 \div \frac{1}{6}$ was written as *ṣaḍ-bhâga-bhâga*,[2] *i.e.*, "one-sixth *bhâga-bhâga*" or "one divided by one-sixth."[3]

(6)[4] *Bhâga-mâtṛ*, *i.e.*, combinations of forms enumerated above. Mahâvîra remarks that there can be twenty-six variations of this type.[5] The following example is given by Srîdhara.[6]

"What is the result when half, one-fourth of one-fourth, one divided by one-third, half plus half of itself, and one-third diminished by half of itself, are added together?"

In modern notation this is

$$\tfrac{1}{2}+(\tfrac{1}{4} \text{ of } \tfrac{1}{4})+(1\div\tfrac{1}{3})+(\tfrac{1}{2}+\tfrac{1}{2} \text{ of } \tfrac{1}{2})+(\tfrac{1}{3}-\tfrac{1}{2} \text{ of } \tfrac{1}{3}).$$

In the old Hindu notation it was written as

1	1	1	1	1	1
2	4	4	1	2	3
			3	1	·1
				2	2

[1] It is only in the Bakhshâlî Manuscript that the term *bhâ* is sometimes placed before or after the quantity affected.

[2] *Cf.* *Triś*, p. 11.

[3] *GSS*, p. 41 (112) gives $2\div\frac{3}{4}$ as *tripâda bhaktaṃ dvikaṃ*, i.e., "two divided by three-fourths."

[4] *Triś*, p. 12; *GSS*, p. 45 (138).

[5] As there are five primary classes enumerated by Mahâvîra, so the total number of combinations is

$$^5C_2+{}^5C_3+{}^5C_4+{}^5C_5=26.$$

[6] *Triś*, p. 12.

The defect of the notation is obvious: $\begin{array}{|c|c|}\hline 1 & 1 \\ 4 & 4 \\ \hline\end{array}$ can be read also as $\frac{1}{4}+\frac{1}{4}$, and $\begin{array}{|c|}\hline 1 \\ 1 \\ 3 \\ \hline\end{array}$ can be read also as $1\frac{1}{3}$; so that the exact meaning of the notation can be understood only by a reference to the question.

The rules for the reduction of the first two classes are those of addition or subtraction, and multiplication. The rule for the reduction of the third and fourth classes (from *ii*) are given together by Brahmagupta:

"The (upper) denominator is multiplied by the denominator and the upper numerator by the same (denominator) increased or diminished by its own numerator."[1]

The rule for *bhâgânubandha* is given by Srîdhara[2] as follows:

(*i*) "In *bhâgânubandha*, add the numerator to the product of the whole number and the denominator."

(*ii*) "Multiply the denominator by the lower denominator and (then) the numerator by the same lower denominator increased by its own numerator."

Other writers give similar rules for reduction in the case of *bhâgânubandha*.

The following example[3] will explain the process of working:

[1] *BrSpSi*, p. 176. The reduction of the form $a \pm \frac{b}{c}$ has been given separately (p. 173).

[2] *Triś*, p. 10. Rule (*i*) is for the reduction of $a+\frac{b}{c}$ and rule (*ii*) is for the reduction of the form

$$\frac{a}{b}+\frac{c}{d} \text{ of } \frac{a}{b}+\frac{e}{f} \text{ of } \left(\frac{a}{b}+\frac{c}{d} \text{ of } \frac{a}{b}\right).$$

[3] *Triś*, p. 11.

Reduce to a proper fraction:

$$3\tfrac{1}{2}+\tfrac{1}{4} \text{ of } 3\tfrac{1}{2}+\tfrac{1}{6} \text{ of } (3\tfrac{1}{2}+\tfrac{1}{4} \text{ of } 3\tfrac{1}{2})+\tfrac{1}{2}+\tfrac{1}{3} \text{ of } \tfrac{1}{2}+\tfrac{1}{4} \text{ of } (\tfrac{1}{2}+\tfrac{1}{3} \text{ of } \tfrac{1}{2}).$$

This was written as

<table>
<tr><td>3
1
2</td><td>1
2</td></tr>
<tr><td>1
4</td><td>1
3</td></tr>
<tr><td>1
6</td><td>1
4</td></tr>
</table>

Adding denominators to numerators of the lower fractions, and applying rule (*i*) to left-hand top compartment to reduce it to a proper fraction, we get

<table>
<tr><td>7
2</td><td>1
2</td></tr>
<tr><td>5
4</td><td>4
3</td></tr>
<tr><td>7
6</td><td>5
4</td></tr>
</table>

Now performing multiplication as directed, *i.e.*, multiplying the denominator of the first fraction by all the lower denominators and the numerator by the sum of the numerators and denominators of the lower fractions, we get

$$\tfrac{7}{2}\times\tfrac{5}{4}\times\tfrac{7}{6}=\tfrac{245}{48}, \text{ and } \tfrac{1}{2}\times\tfrac{4}{3}\times\tfrac{5}{4}=\tfrac{20}{24}$$

i.e.,

<table>
<tr><td>245
48</td><td>20
24</td></tr>
</table>

Then making denominators similar (*savarṇana*), we have

<table>
<tr><td>245
48</td><td>40
48</td></tr>
</table>

;

performing the addition we have $\frac{285}{48}$ or $5\frac{5}{6}$ as the result.

The rule for *bhâgâpavâha* is given in all the works on *pâṭîgaṇita*. It is the same as that for *bhâgânubandha*, except that "addition" or "increase" is replaced by "subtraction" or "decrease" in the enunciation of the rule for *bhâgâpavâha*.

Lowest Common Multiple. Mahâvîra[1] was the first amongst the Indian mathematicians to speak of the lowest common multiple in order to shorten the process. He defines *niruddha* (L. C. M.) as follows:

"The product of the common factors of the denominators and their resulting quotients is called *niruddha*."

The process of reducing fractions to equal denominators is thus described by him:[2]

"The (new) numerators and denominators, obtained as products of multiplication of (each original) numerator and denominator by the (quotient of the) *niruddha* (*i.e.*, L. C. M.) divided by the denominator give fractions with the same denominator."

Bhâskara II[3] does not mention *niruddha* but observes that the process can be shortened. He says:

"The numerator and denominator may be multiplied by the intelligent calculator by the other denominator abridged by the common factor."

The Eight Operations. Operations with fractions were known in India from very early times, the method of performing them being the same as now.

[1] *GSS*, p. 33 (56).
[2] *GSS*, p. 33 (56).
[3] *L*, p. 6.

Although Aryabhaṭa does not mention the elementary operations, there is evidence to show that he knew the method of division by fraction by inverting it. All the operations are found in the Bakhshâlî Manuscript (*c.* 200).

Addition and Subtraction. These operations were performed after the fractions were reduced to a common denominator. Thus Srîdhara says:[1]

"Reduce the fractions to a common denominator and then add the numerators. The denominator of a whole number is unity."

Brahmagupta and Mahâvîra give the method under *Bhâgajâti*. Mahâvîra differs from other writers in giving the methods of the summation of arithmetic and geometric series under the title of addition (*saṁkalita*).[2] Later writers follow Srîdhara.

Multiplication. Brahmagupta says:[3]

"The product of the numerators divided by the product of the denominators is the (result of) multiplication of two or more fractions."

While all other writers give the rule in the same way as Brahmagupta, Mahâvîra refers to cross reduction in order to shorten the work:[4]

"In the multiplication of fractions, the numerators are to be multiplied by the numerators and the denominators by denominators, after carrying out the process of cross reduction,[5] if that be possible."

[1] *Triś*, p. 7.
[2] *Cf. GSS*, pp. 28 (22) ff.
[3] *BrSpSi*, p. 173.
[4] *GSS*, p. 25 (2).
[5] *Vajrâpavartana-vidhi*, i.e., "cancellation crosswise," thus

$\frac{2}{3} \times \frac{6}{4} =$ $= \frac{2}{2} = 1.$

Division. Although the elementary operations are not mentioned in the *Aryabhaṭîya*, the method of division by fraction is indicated under the Rule of Three. The Rule of Three states the result as $\frac{f \times i}{p}$.[1] When these quantities are fractional, we get an expression of the form $\frac{\frac{a}{b} \times \frac{c}{d}}{\frac{m}{n}}$, for the evaluation of which Āryabhaṭa I states:

"The multipliers and the divisor are multiplied by the denominators of each other."

As will be explained later on (p. 204) the quantities are written as

$\frac{a}{b}$	$\frac{m}{n}$
$\frac{c}{d}$	

Transferring the denominators we have

a n c	m b d

Performing multiplication, the result is $\frac{anc}{mbd}$.

The above interpretation of a rather obscure line[2] in the *Aryabhaṭîya* is based on the commentaries of Sûryadeva and Bhâskara I. Thus Sûryadeva says:

[1] Where f = *phala*, *i.e.*, "fruit," i = *icchâ*, *i.e.*, "demand or requisition" p'=*pramâṇa*, *i.e.*, "argument."

[2] *Â*, p. 43. Previous writers seem to have been misled by the commentary of Parameśvara which is very vague; *cf*; Clark (p. 40) and P. C. Sengupta (p. 25).

"Here by the word *guṇakâra* is meant the multiplier and multiplicand, *i.e.*, the *phala* and *icchâ* quantities that are multiplied together. By *bhâgahâra* is meant the *pramâṇa* quantity. The denominators of the *phala* and *icchâ* are taken to the *pramâṇa*. The denominator of the *pramâṇa* is taken with the *phala* and *icchâ*. Then multiplying these, *i.e.*, (the numerators of) the *phala* and *icchâ* and this denominator, and dividing by (the product of) the numbers standing with the *pramâṇa*, the result is the quotient of the fractions."

Brahmagupta[1] gives the method of division as follows:

"The denominator and numerator of the divisor having been interchanged, the denominator of the dividend is multiplied by the (new) denominator and its numerator by the (new) numerator. Thus division of *proper fractions* is performed."

Śrîdhara[2] adds the following to the method of multiplication:

"Having interchanged the numerator and denominator of the divisor, the operation is the same as before."[3]

Mahâvîra[4] explains the method thus:

"After having made the numerator of the divisor[5] its denominator (and *vice versa*) the operation is the same as in multiplication."

"Or,[6] when (the fractions constituting) the divisor

[1] *BrSpSi*, p. 173.

[2] *Triś*, p. 8.

[3] *i.e.*, the same as that of multiplication.

[4] *GSS*, p. 26 (8).

[5] Mahâvîra uses the term *pramâṇa-râśi* for divisor, showing thereby its connection with the 'rule of three.'

[6] This is similar to the way in which Âryabhaṭa I expresses the method.

and dividend are multiplied by the denominators of each other and these products are without denominators, (the operation) is as in the division of whole[1] numbers."

Square and Square-root. Brahmagupta[2] says:

"The square of the numerator of a proper fraction divided by the square of the denominator gives the square."

"The square-root of the numerator of a proper fraction divided by the square-root of the denominator gives the square-root."

Other works contain the same rules.

Cube and Cube-root. Śrîdhara[3] gives the rule as follows:

"The cube of the numerator divided by the cube of the denominator gives the cube, and the cube-root of the numerator divided by the cube-root of the denominator gives the cube-root."

Other works give the same rules.

Unit Fractions. Mahâvîra has given a number of rules for expressing any fraction as the sum of a number of unit fractions.[4] These rules do not occur in any other work, probably because they were not considered important or useful.

(1) *To express* 1 *as the sum of a number* (n) *of unit fractions.*

The rule for this is:[5]

"When the sum of the different quantities having

[1] The term for whole number is *sakala.*

[2] *BrSpSi*, p. 174.

[3] *Triś*, p. 9.

[4] There is no technical term for unit fraction. The term used is *rûpâṁśaka-râśi*, *i.e.*, "quantity with one as numerator."

[5] *GSS*, p. 36 (75).

one for their numerator is 1, the (required) denominators are such as, beginning with 1, are in order multiplied by 3, the first and the last being multiplied again by 2 and $\frac{2}{3}$."

Algebraically the rule is

$$1 = \frac{1}{2} + \frac{1}{3} + \frac{1}{3^2} + \frac{1}{3^3} + \dots + \frac{1}{3^{n-2}} + \frac{1}{2.3^{n-2}}.$$

(2) *To express* 1 *as the sum of an odd number of unit fractions.*

The rule for this is stated thus:[1]

"When the sum of the quantities (fractions) having one for each of their numerators is one, the denominators are such as, beginning with two, go on rising in value by one, each being further multiplied by that which is (immediately) next to it and then halved."

Algebraically this is

$$1 = \frac{1}{2.3.\frac{1}{2}} + \frac{1}{3.4.\frac{1}{2}} + \dots + \frac{1}{(2n-1).2n.\frac{1}{2}} + \frac{1}{2n.\frac{1}{2}}$$

(3) *To express a unit fraction as the sum of a number of other fractions, the numerators being given.*[2]

The rule for this is:

"The denominator of the first (of the supposed or given numerators) is the denominator of the sum, that of the next is this combined with its numerator and so on; and then multiply (each denominator) by that which is next to it, the last being multiplied by its own numerator. (This gives the required denominators)."

[1] *GSS*, p. 36(77).

[2] Each may be one. *GSS*, p. 36(78).

Algebraically this gives:

$$\frac{1}{n}=\frac{a_1}{n\,(n+a_1)}+\frac{a_2}{(n+a_1)(n+a_1+a_2)}+\ldots.$$

$$+\frac{a_{r-1}}{(n+a_1+a_2+\ldots.+a_{r-2})(n+a_1+a_2+\ldots.+a_{r-1})}$$

$$+\frac{a_r}{a_r(n+a_1+a_2+\ldots.+a_{r-1})}$$

By taking $a_1=a_2=\ldots=a_r=1$, we get unit fractions. When these are not unity, the fractions may not be in their lowest terms.

(4) *To express any fraction as the sum of unit fractions.*

The rule is:[1]

"The denominator (of the given fraction) when combined with an optionally chosen number and then divided by the numerator so as to leave no remainder, becomes the denominator of the first numerator (which is one); the optionally chosen quantity when divided by this and by the denominator of the sum is the remainder. To this remainder the same process is applied."

Let the number i be so chosen that $\frac{q+i}{p}$ is an integer $= r$; then the rule gives

$$\frac{p}{q}=\frac{1}{r}+\frac{i}{r.q}$$

of which the first is a unit fraction and a similar process can be employed to the remainder to get other unit fractions. In this case the result depends upon the optionally chosen quantities.

[1] *GSS*, p. 37(80).

(5) *To express a unit fraction as the sum of two other unit fractions.*

The following two rules are given:[1]

(*i*) "The denominator of the given sum multiplied by a properly chosen number is the (first) denominator, and this divided by the previously chosen number minus one gives the other; or (*ii*) the two denominators are the factors[2] of the denominator of the sum, each multiplied by their sum."

Expressed algebraically the rules are:

$$(i) \quad \frac{1}{n} = \frac{1}{p.n} + \frac{1}{\frac{p.n}{p-1}}\ ^{3}$$

$$(ii) \quad \frac{1}{a.b} = \frac{1}{a(a+b)} + \frac{1}{b(a+b)}$$

(6) *To express any fraction as the sum of two other fractions whose numerators are given.*

The rule for this is:[4]

"Either numerator multiplied by a chosen number, then combined with the other numerator, then divided by the numerator of the sum so as to leave no remainder, and then divided by the chosen number and multiplied by the denominator of the sum gives rise to one denominator. The denominator corresponding to the other (numerator), however, is this (denominator) multiplied by the chosen quantity."

[1] *GSS*, p. 37(85).

[2] *hâra-hâra-labdha*, lit. "the divisor and quotient by that divisor."

[3] The integer p is so chosen that n is divisible by $(p-1)$.

[4] *GSS*, p. 38(87).

Algebraically the rule is

$$\frac{m}{n}=\frac{a}{\frac{ap+b}{m}\times\frac{n}{p}}+\frac{b}{\frac{ap+b}{m}\times\frac{n}{p}\times p}\ ^{1}$$

A particular[1] case of this would be

$$\frac{m}{n}=\frac{a}{\frac{an+b}{m}}+\frac{b}{\frac{an+b}{m}\times n}$$

provided that $(an+b)$ is divisible by m.

(7) *To express a given fraction as the sum of an even number of fractions whose numerators are previously assigned.*

The rule for this is:[2]

"After splitting up the sum into as many parts, having one for each of their numerators, as there are pairs (among the given numerators), these parts are taken as the sum of the pairs, and (then) the denominators are found according to the rule for finding two fractions equal to a given unit fraction."

12. THE RULE OF THREE

Terminology. The Hindu name for the Rule of Three terms is *trairâśika* ("three terms," hence "the rule of three terms"). The term *trairâśika* can be traced back to the beginning of the Christian era as it occurs in the

[1] Evidently, the chosen number p must be a divisor of n, and such that $\frac{ap+b}{m}$ is an integer.

The solution given does not hold for any values of a and b, but only for such values as allow of an integer p to be so chosen as to satisfy the required conditions.

[2] *GSS*, p. 38(89).

Bakhshâlî Manuscript,[1] in the *Âryabhaṭîya* and in all other works on mathematics. About the origin of the name Bhâskara I (*c.* 525) remarks:[2] "Here three quantities are needed (in the statement and calculation) so the method is called *trairâśika* ("the rule of three terms")."

A problem on the rule of three has the form:

If p yields f, what will i yield?

In the above, the three terms are p, f and i. The Hindus called the term p, *pramâṇa* ("argument"), the term f, *phala* ("fruit") and the term i, *icchâ* ("requisition"). These names are found in all the mathematical treatises. Sometimes they are referred to simply as the first, second and third respectively. Âryabhaṭa II differs from other writers in giving the names *mâna*, *vinimaya* and *icchâ* respectively to the three terms. It has been pointed out by most of the writers that the first and third terms are similar, *i.e.*, of the same denomination.

The Method. Âryabhaṭa I (499) gives the following rule for solving problems on the Rule of Three:

"In the Rule of Three, the *phala* ("fruit"), being multiplied by the *icchâ* ("requisition") is divided by the *pramâṇa* ("argument"). The quotient is the fruit corresponding to the *icchâ*. The denominators of one being multiplied with the other give the multiplier (*i.e.*, numerator) and the divisor (*i.e.*, denominator)."[3]

[1] The term *râśi* is used in the enumeration of topics of mathematics in the *Sthânâṅga-sûtra* (*c.* 300 B.C.) (*Sûtra* 747). There it probably refers to the Rules of Three, Five, Seven, etc.

[2] In his commentary on the *Aryabhaṭîya*.

[3] The above corresponds to *âryâ* 26 and the first half of *âryâ* 27 of the *Gaṇitapâda* of the *Aryabhaṭîya*; compare the working of *Example* I, where the interchange of denominators takes place. See also pp. 195*f*.

Brahmagupta gives the rule thus:

"In the Rule of Three *pramâṇa* ("argument"), *phala* ("fruit") and *icchâ* ("requisition") are the (given) terms; the first and the last terms must be similar. The *icchâ* multiplied by the *phala* and divided by the *pramâṇa* gives the fruit (of the demand)."[1]

Śrîdhara states:

"Of the three quantities, the *pramâṇa* ("argument") and *icchâ* ("requisition") which are of the same denomination are the first and the last; the *phala* ("fruit") which is of a different denomination stands in the middle; the product of this and the last is to be divided by the first."[2]

Mahâvîra writes:

"In the Rule of Three, the *icchâ* ("requisition") and the *pramâṇa* ("argument") being similar, the result is the product of the *phala* and *icchâ* divided by the *pramâṇa*."[3]

Âryabhaṭa II introduces a slight variation in the terminology. He says:

"The first term is called *mâna*, the middle term *vinimaya* and the last one *icchâ*. The first and the last are of the same denomination. The last multiplied by the middle and divided by the first gives the result."[4]

Bhâskara II, Nârâyaṇa and others give the rule in the same form as Brahmagupta or Śrîdhara.

The Hindu method of working the rule may be illustrated by the following examples taken from the *Triśatikâ*:

[1] *BrSpSi*, p. 178.
[2] *Triś*, p. 15.
[3] *GSS*, p. 58(2).
[4] *MSi*, p. 149.

Example I.[1] "If one *pala* and one *karṣa* of sandal wood are obtained for ten and a half *pana*, for how much will be obtained nine *pala* and one *karṣa*?"

Here 1 *pala* and 1 *karṣa*$=1\frac{1}{4}$ *pala*, and 9 *pala* and 1 *karṣa*$=9\frac{1}{4}$ *pala* are the similar quantities. The "fruit" $10\frac{1}{2}$ *paṇa* corresponding to the first quantity ($1\frac{1}{4}$ *pala*) is given, so that

pramâṇa (argument) $= 1\frac{1}{4}$
phala (fruit) $= 10\frac{1}{2}$
icchâ (requisition) $= 9\frac{1}{4}$

The above quantities are placed in order as

$$\begin{array}{|c|c|c|} 1 & 10 & 9 \\ 1 & 1 & 1 \\ 4 & 2 & 4 \end{array}$$

Converting these into proper fractions we have

$$\begin{array}{|c|c|c|} 5 & 21 & 37 \\ 4 & 2 & 4 \end{array}$$

Multiplying the second and the last and dividing by the first, we have

$$\begin{array}{|c|c|} \begin{matrix}21\\2\end{matrix} & \begin{matrix}5\\4\end{matrix} \\ \hline \begin{matrix}37\\4\end{matrix} & \end{array} = \frac{\frac{21}{2}\times\frac{37}{4}}{\frac{5}{4}}$$

Or transferring denominators $\begin{array}{|c|c|} 21 & 5 \\ 4 & 2 \\ 37 & 4 \end{array} = \dfrac{21.4.37}{5.2.4}$ *pala*

$=$ 4 *purâṇa*, 13 *paṇa*, 2 *kâkiṇî* and 16 *varâṭaka*.

In actual working the intermediate step

$$\frac{\frac{21}{2}\times\frac{37}{4}}{\frac{5}{4}}$$

[1] *Triś*, p. 15.

was not written. The denominators of the multipliers were transferred to the side of the divisor and that of the divisor to the multipliers, thus giving at once

$$\frac{21.4.37}{5.2.4}.$$

Example II.[1] "Out of twenty necklaces each of which contains eight pearls, how many necklaces, each containing six pearls, can be made?"

Firstly, we have

1	8	20

The result (performing the operation of the Rule of Three) is 160 pearls.

Secondly, perform the operation of the Rule of Three on the following:

If 6 pearls are contained in one necklace, how many necklaces will contain 160 pearls?

Placing the numbers, we have

6	1	160

Result: necklaces 26, part of necklace $\boxed{\begin{matrix}2\\3\end{matrix}}$.

Inverse Rule of Three. The Hindu name for the Inverse Rule of Three is *vyasta-trairâśika* (lit. "inverse rule of three terms"). After describing the method of the Rule of Three the Hindu writers remark that the operation should be reversed when the proportion is inverse. Thus Śrîdhara observes:

"The method is to multiply the middle term by the first and to divide by the last, in case the proportion is different."[2]

[1] *Triś*, p. 17.
[2] *Triś*, p. 18.

Mahâvîra says:

"In the case of this (proportion) being inverse, the operation is reversed."[1]

Bhâskara II writes:

"In the inverse (proportion), the operation is reversed."[2]

He further observes:

"Where with increase of the *icchâ* (requisition) the *phala* decreases or with its decrease the *phala* increases, there the experts in calculation know the method to be the Inverse Rule of Three."[3]

"Where the value of living beings is regulated by their age; and in the case of gold, where the weight and touch are compared; or when heaps are subdivided,[4] let the Inverse Rule of Three be used."[5]

Example: Example II given under the Rule of Three above has been solved also by the application of the Inverse Rule as follows:

"Statement

8	20	6

Result: necklaces

26
2
3

"

Here if the *icchâ*, i.e., the number of pearls in a necklace, increases, the *phala*, *i.e.*, the number of necklaces, decreases, so that the Inverse Rule of Three is applied.

Appreciation of the Rule of Three. The Rule of Three was highly appreciated by the Hindus because

[1] *GSS*, p. 58(2).
[2] *L*, p. 17.
[3] *L*, p. 17.
[4] "When heaps of grain, which have been meted with a small measure, are again meted with a larger one, the number decreases ...." (Com. of Sûryadâsa).
[5] *L*, p. 18.

of its simplicity and its universal application to ordinary problems. The method as evolved by the Hindus gives a ready rule which can be applied even by the "ignorant person" to solve problems involving proportion, without fear of committing errors. Varâhamihira (505) writes:

"If the sun performs one complete revolution in a year, how much does he accomplish in a given number of days? Does not even an ignorant person calculate the sun in such problems by simply scribbling with a piece of chalk?"[1]

Bhâskara II has eulogised the method highly at several places in his work. His remarks are:

"The Rule of Three is indeed, (the essence of) arithmetic."[2]

"As Lord Śri Nârâyaṇa, who relieves the sufferings of birth and death, who is the sole primary cause of the creation of the universe, pervades this universe through His own manifestations as worlds, paradises, mountains, rivers, gods, men, demons, etc., so does the Rule of Three pervade the whole of the science of calculation. ....Whatever is computed whether in algebra or in, this (arithmetic) by means of multiplication and division may be comprehended by the sagacious learned as the Rule of Three. What has been composed by the sages through the multifarious methods and operations such as miscellaneous rules, etc., teaching its easy variations, is simply with the object of increasing the comprehension of the duller intellects like ourselves."[3]

On another occasion Bhâskara II observes:

[1] *PSi*, iv. 37.

[2] *L*, p. 15. The same remark occurs in *SiŚi*, *Golâdhyâya*, *Praśnâdhyâya*, verse 3.

[3] *L*, p. 76.

"Leaving squaring, square-root, cubing and cube-root, whatever is calculated is certainly variation of the Rule of Three, nothing else. For increasing the comprehension of duller intellects like ours, what has been written in various ways by the learned sages having loving hearts like that of the bird *cakora*, has become arithmetic."[1]

Proportion in the West. The history of the Hindu rules of proportion shows how much the West was indebted to India for its mathematics. The Rule of Three occurs in the treatises of the Arabs and mediæval Latin writers, where the Hindu name 'Rule of Three' has been adopted. Although the Hindu names of the terms were discarded, the method of placing the terms in a line, and arranging them so that the first and last were similar, was adopted. Thus Digges (1572) remarked,[2] "Worke by the Rule ensueing ...... Multiplie the last number by the seconde, and diuide the Product by the first number," ... "In the placing of the three numbers this must be observed, that the first and third be of one Denomination." The rule, as has been already stated, was perfected in India in the early centuries of the Christian era. It was transmitted to the Arabs probably in the eighth century and thence travelled to Europe, where it was held in very high esteem[3] and called the "Golden Rule."

Compound Proportion. The Hindu names for compound proportion are the Rule of Five, the Rule of

[1] *SiŚi*, *Golâdhyâya*, *Praśnâdhyâya*, verse 4.

[2] Quoted by Smith, *l.c.* p. 488.

[3] The Arabs, too, held the method in very high esteem as is evidenced by Al-Bîrûnî's writing a separate treatise, *Fi raśikat al-hind* ("On the *raśika* of the Hindus") dealing with the Hindu Rules of Three or more terms. Compare also *India* (I. 313) where an example of *vyasta-trairâśika*, ("the Inverse Rule of Three") is given.

Seven, the Rule of Nine, etc., according to the number of terms involved in the problems. These are sometimes grouped under the general appellation of the "Rule of Odd Terms." The above technical terms as well as the rules were well-known in the time of Âryabhaṭa I (499), although he mentions the Rule of Three only. That the distinction between the Rule of Three and Compound Proportion is more artificial than real was stressed by Bhâskara I (*c.* 525) in his commentary on the *Âryabhaṭîya*. He says:

"Here Âcârya Âryabhaṭa has described the Rule of Three only. How the well-known Rules of Five, etc. are to be obtained? I say thus: The Âcârya has described only the fundamentals of *anupâta* (proportion). All others such as the Rule of Five, etc., follow from that fundamental rule of proportion. How? The Rule of Five, etc., consist of combinations of the Rule of Three .... In the Rule of Five there are two Rules of Three, in the Rule of Seven, three Rules of Three, and so on. This I shall point out in the examples."

Remarks similar to the above concerning the Rules of Five, Seven, etc., have been made by the commentators of the *Lîlâvatî*, especially by Gaṇeśa and Sûryadâsa.[1]

In problems on Compound Proportion, two sets of terms are given. The first set which is complete is called *pramâṇa pakṣa* (argument side) and the second set in which one term is lacking is called the *icchâ pakṣa* (requisition side).

The Method. The rule relating to the solution of problems in compound proportion has been given by Brahmagupta as follows:

"In the case of odd terms beginning with three

[1] Noted by Colebrooke, *l.c.*, p. 35, note.

terms[1] upto eleven, the result is obtained by transposing the fruits of both sides, from one side to the other, and then dividing the product of the larger set of terms by the product of the smaller set. In all the fractions the transposition of denominators, in like manner, takes place on both sides."[2]

Srîdhara says:

"Transpose the two fruits from one side to the other, then having transposed the denominators (also in like manner) and multiplied the numbers (so obtained on each side), divide the side with the larger number of terms by the other (side)."[3]

Mahâvîra[4] and Âryabhaṭa II[5] have given the rule in the same way as Srîdhara. Bhâskara II has given it thus:

"In the rules of five, seven, nine or more terms, after having taken the *phala* (fruit) and *chid*[6] from its

[1] It should be observed that, as stated above, the Rule of Three is a particular case of the above Rule of Odd terms. Brahmagupta is the only Hindu writer to have included the Rule of Three also in the above rule. Some Arab writers have followed him in this respect by not writing the terms of the Rule of Three in a line, but arranging them in compartments, as for the other rules of odd terms.

[2] *BrSpSi*, p. 178.

[3] *Triś*, p. 19.

[4] *GSS*, p. 62 (32).

[5] *MSi*, p. 150, rules 26 and 27 (repeated with a slight variation).

[6] The commentators differ as regards the interpretation of this word. Some take it to mean "divisor," *i.e.*, "denominator," while others say that it means "the fruit of the other side." The rule is, however, correct with either interpretation. The first interpretation, however, brings Bhâskara's version in line with those of his predecessors. It may be mentioned here that Ârya-bhaṭa II repeats the rule twice. At first he does not direct the transposition of denominator, and at the second time he does so.

own side to the other, the product of the larger set of terms divided by the product of the smaller set, gives the result (or produce sought)."[1]

Illustration. We shall illustrate the Hindu method of working by solving the following example taken from the *Lîlâvatî*:

"If the interest of a hundred in one month be five, what will be the interest of 16 in 12 months? Also find the time knowing the interest and principal; and tell the principal knowing the time and interest."

To find interest.

The first set of terms (*pramâṇa pakṣa*) is:

100 *niṣka*, 1 month, 5 *niṣka* (*phala*)

The second set (*icchâ pakṣa*) is:

16 *niṣka*, 12 months, x *niṣka*

The terms are now written in compartments[2] as below:

100	16
1	12
5	0

[3]

[1] *L*, p. 18.

[2] The terms of the same denomination are written in compartments in the same horizontal line.

[3] The figures are written in compartments in order to facilitate the writing of fractions and also to denote the side which contains more terms after transposition of fruits. Sometimes, the compartment corresponding to an absent term is left vacant as we find in a copy of Munîśvara's *Pâṭîsâra* (in the Government Sanskrit Library at Benares). When the terms are written in compartments, the symbol o to denote the unknown or absence of a term is unnecessary. In some commentaries on the *Lîlâvatî* (Asiatic Society of Bengal manuscripts) we find the numbers written without compartments, but in such cases the symbol o is used to denote the absence of a term. After transposition, the side on which o occurs contains a smaller number of terms than the other.

In the above 5 (written lowest) is the "fruit" of the first side, and there is no "fruit" on the second side. Interchanging the fruits we get

100	16
1	12
0	5

The larger set of terms is on the second "side." The product of the numbers is 960. The product of the numbers on the side of the smaller set of terms is 100. Therefore, the required result is $\frac{960}{100}=\frac{48}{5}$, written as $\left|\begin{smallmatrix}48\\5\end{smallmatrix}\right|$ or 9 *niṣka*, fraction $\left|\begin{smallmatrix}3\\5\end{smallmatrix}\right|$

To find Time:

Here the sides are

100 *niṣka*, 1 month, 5 *niṣka*

and 16 *niṣka*, x months, $\frac{48}{5}$ *niṣka*

The terms are written as

100	16
1	0
5	48 5

Transposing the fruits, *i.e.*, transposing the numbers in the bottom compartment, we get

100	16
1	0
48 5	5

Transposing the denominators we have

100	16
1	0
48	5 5

Here, the larger set of terms is on the first side and their product is 4800. The product of the numbers on the side of the smaller set is 400. Therefore, the result is

$$\begin{array}{|r|} 4800 \\ 400 \\ \hline \end{array} = \frac{4800}{400} = 12 \text{ months.}$$

To know the principal:

The first side is

100 *niṣka*, 1 month, 5 *niṣka*

The second side is

x *niṣka*, 12 months, $\frac{48}{5}$ *niṣka*

This is written as

100	0
1	12
5	48 5

After transposition of fruits (*i.e.*, the terms in the bottom cells) we have

100	0
1	12
48 5	5

Transposing denominators we get

100	0
1	12
48	5 5

The product of the numbers in the larger set divided by the product of the numbers in the smaller set, gives

$$\frac{4800}{300} = 16 \text{ } \textit{niṣka}.$$

Rule of Three as a Particular Case. According to Brahmagupta, the above method may be applied to the Rule of Three. Taking the first example solved under the Rule of Three, above, and placing the terms we have

21	0
2	
5	37
4	4

[1]

Transposing the fruits, we have

21	0
2	
37	5
4	4

Transposing denominators, we get

21	0
	2
37	5
4	4

Therefore, the result is $\dfrac{21.37.4}{2.5.4}$ as before.

If we consider the term corresponding to the unknown as the fruit, the terms should be set as below:

[1] Here, we consider $\frac{5}{4}$ *pala* of sandal wood as the "fruit" of $\frac{21}{2}$ *paṇa* (money). The previous method forces us to consider $\frac{21}{2}$ *paṇa* as the "fruit" or the middle term, because the "first" and "third" are directed to be alike. It will be observed that any of the terms may be considered to be the fruit in the alternative method given here.

5	37
4	4
21	
2	o

Hence, as before,[1] the result is $\frac{37.4.21}{5.4.2}$

The above method of working the Rule of Three is found among the Arabs,[2] although it does not seem to have been used in India after Brahmagupta. This points to the indebtedness of the Arabs to Brahmagupta especially, for their knowledge of Hindu arithmetic.

Written as above the method of working the Rule of Three appears to be the same as the method of proportion. In the same way the rule of other odd terms, when properly translated into modern symbolism, is nothing but the method of proportion. It has been stated by Smith[3] that the Hindu methods of solution "fail to recognize the relation between the Rule of Three and proportion." This statement appears to have been made without sufficient justification, for the solutions have been evidently obtained by the use of the ideas of proportionality and variation. The aim of the Hindu works is to give a method which can be readily used by common people. For this very reason, the cases in which the variation is inverse have been enumerated. Considered as a method which stimulated the student to think for himself, the method is certainly

[1] The product of the numbers on the side of the larger set is divided by the product of the numbers on the side of the smaller set. o in this case is not a number. It is the symbol for the unknown or absence.

[2] Thus Rabbi ben Ezra wrote $\frac{47}{7}\ \frac{63}{0}$ for $47 : 7 = 63 : x$. See Smith, *l.c.*, p. 489f.

[3] *l.c.*, p. 488.

defective, but for practical purposes, it is, in our opinion, the best that could be devised.

13. COMMERCIAL PROBLEMS

Interest in Ancient India. The custom of taking interest is a very old one. In India it can be definitely traced back to the time of Pâṇini (*c.* 700 B.C.) who in his Grammar lays down rules validating the use of the suffix *ka* to number names in case of "an interest, a rent, a profit, a tax or a bribe given."[1] The interest became due every month and the rate of interest was generally given per hundred,[2] although this was not always the case. The rate of interest varied in different localities and amongst different classes of people, but an interest of fifteen per cent per year seems to have been considered just. Thus in Kauṭilya's *Arthaśâstra*, a work of the fourth century B.C., it is laid down: "an interest of a *paṇa* and a quarter per month per cent is just. Five *paṇa* per month per cent is commercial interest. Ten *paṇa* per month per cent prevails in forests. Twenty *paṇa* per month per cent prevails among sea traders."[3] The *Gotama Sûtra* states: "an interest of five *mâṣâ* per twenty (*kârṣâpaṇa*) is just."[4]

Interest in Hindu Ganita. The ordinary problems relating to the finding out of interest, principal or time etc., the other quantities being given, occur in the section dealing with the Rule of Five. The Hindu

[1] Pâṇini's Grammar, v. i. 22, 47, 49.

[2] It has been pointed out by B. Datta that the idea of *per cent* first originated in India. See his article in the *American Mathematical Monthly*, XXXIV, p. 530.

[3] *Arthaśâstra*, edited and translated into English by R. Shamsastry, Mysore, III, ii, p. 214.

[4] *Gotama Sûtra*, xii. 26. Since 20 *mâṣâ* equal a *kârṣâpaṇa*, the rate is 15 per cent annually.

works generally contain a section called *miśraka-vyavahâra* ("calculations relating to mixed quantities") in which occur miscellaneous problems on interest. The contents of this section vary in different works, according to their size and scope. Thus the *Âryabhaṭîya* contains only one rule relating to a problem on interest, whilst the *Gaṇita-sâra-saṁgraha* has a large number of such rules and problems.

Problem involving a Quadratic Equation. Âryabhaṭa I (499) gives a rule for the solution of the following problem:

The principal sum $p(=100)$ is lent for one month (interest unknown$=x$). This unknown interest is then lent out for $t(=\text{six})$ months. After this period the original interest (x) plus the interest on this interest amounts to $A(=\text{sixteen})$. The rate-interest (x) on the principal (p) is required.

The above problem requires the solution of the quadratic equation

$$tx^2+px-Ap=0,$$

which gives $$x=\frac{-p/2\pm\sqrt{(p/2)^2+Apt}}{t}.$$

The negative value of the radical does not give a solution of the problem; so the result is

$$x=\frac{\sqrt{Apt+(p/2)^2}-p/2}{t}.$$

This is stated by Âryabhaṭa I as follows:

"Multiply the sum of the interest on the principal and the interest on this interest (A) by the time (t) and by the principal (p). Add to this result the square of half the principal $\{(p/2)^2\}$. Take the square-root of this. Subtract half the principal $(p/2)$ and divide the remainder by the time (t). The result will be the (unknown) interest (x) on the principal."[1]

Brahmagupta (628) gives a more general rule. His problem is:

The principal (p) is lent out for t_1 months and the unknown interest on this ($=x$) is lent out for t_2 months at the same rate and becomes A. To find x.

This gives the quadratic

$$x^2 + \frac{pt_1}{t_2}x - \frac{Apt_1}{t_2} = 0,$$

whose solution is

$$x = \pm\sqrt{\frac{Apt_1}{t_2} + \left(\frac{pt_1}{2t_2}\right)^2} - \frac{pt_1}{2t_2}.$$

The negative value of the radical does not give a solution of the problem, so it is discarded.

Brahmagupta states the formula thus:

"Multiply the principal (p) by its time (t_1) and divide by the other time (t_2) (placing the result) at two places. Multiply the first of these by the mixture (A). Add to this the square of half the other. Take the square-root of this (sum). From the result subtract half the other. This will be the interest (x) on the principal."[2]

Other Problems. Mahâvîra (850) gives two other types of problems on "mixture" requiring the solution of simultaneous equations. As an example of the first type may be mentioned the following:[3]

"It has been ascertained that the interest for $1\frac{1}{2}$ months (t=rate-time) on 60 (c=rate-capital) is $2\frac{1}{2}$ (i=

[1] *A*, p. 41. The Sanskrit terms are: *mûla*=principal, *phala* =interest.

[2] *BrSpSi*, p. 183. This rule is also given by Mahâvîra, *GSS*, p. 71 (44).

[3] *GSS*, p. 69(32).

rate-interest). The interest (on the unknown capital P) for an unknown period (T) is 24 ($=I$), and 60 ($=m$ $=P+T$) is the time combined with the capital lent out. What is the time (T) and what is the capital (P)?"

The problem gives:

$$\frac{iPT}{ct} = I \quad \ldots \quad (1)$$

$$P+T = m \quad \ldots \quad (2)$$

$$\therefore \; P-T = \pm\sqrt{m^2 - \frac{ct}{i} \times 4I}$$

Hence $$P = \tfrac{1}{2}\left(m \pm \sqrt{m^2 - \frac{ct}{i} \times 4I}\right),$$

and $$T = \tfrac{1}{2}\left(m \mp \sqrt{m^2 - \frac{ct}{i} \times 4I}\right)$$

The above result is stated by Mahâvîra thus:

"From the square of the mixture (m) subtract the rate-capital (c) divided by the rate-interest (i) multiplied by the rate-time (t) and four times the given interest (4I). Then the operation of *saṅkramaṇa*[1] is performed in relation to the square-root of this and the mixture (m)."[2]

The second type of problems may be illustrated by the following example:

"The interest on 30 (P) is 5 (I) for an unknown

[1] Given the numbers a and b, the process of *saṅkramaṇa* is the finding out of half their sum and difference *i.e.* $\frac{a+b}{2}$ and $\frac{a-b}{2}$.

[2] *GSS*, p. 68(29). It should be noted that both the signs of the radical are used.

period (T), and at an unknown rate of interest (i) per 100 (c) per $1\frac{1}{2}$ month (t). The mixture ($m=i+T$) is $12\frac{1}{2}$. Find i and T."[1]

The solution is given by

$$T = \tfrac{1}{2}\left(m \pm \sqrt{m^2 - \frac{ctI.4}{P}}\right),$$

and consequently

$$i = \tfrac{1}{2}\left(m \mp \sqrt{m^2 - \frac{ctI.4}{P}}\right).$$

Mahâvîra states the solution thus:

"The rate-capital (c) multiplied by its time (t) and the interest (I) and the square of two (=4) is divided by the other capital (P). Then perform the operation of *saṅkramaṇa* in relation to the square-root of the remainder (obtained as the result of subtracting the quotient so obtained) from the square of the mixture (m) and the mixture."[2]

Miscellaneous Problems on Interest. Besides the problems given above various other interesting problems are found in the Hindu works on *pâṭigaṇita*. Thus Brahmagupta gives the solution of the following problem:

Example. In what time will a given sum s, the interest on which for t months is r, become k times itself?

The rule for the solution of the above is:[3]

"The given sum[4] multiplied by its time and divided

[1] *GSS*, p. 69(34).
[2] *GSS*, p. 69(33).
[3] *BrSpSi*, p. 181.
[4] The Sanskrit term used is *pramâṇa* (argument).

by the interest,[1] being multiplied by the factor[2] less one, is the time (required)."

The *Gaṇita-sâra-saṁgraha* (850) contains a large number of problems relating to interest. Of these may be mentioned the following:

(1) "In this (problem), the (given) capitals are ($c_1=$) 40, ($c_2=$) 30, ($c_3=$) 20 and ($c_4=$) 50; and the months are ($t_1=$) 5, ($t_2=$) 4, ($t_3=$) 3 and ($t_4=$) 6 (respectively). The sum of the interests is ($m=$) 34. (Assuming the rate of interest to be the same in each case, find the amounts of interest in each case)."[3]

Here, if the rate of interest per month for 1 be r, then

$$r = \frac{x_1}{c_1t_1} = \frac{x_2}{c_2t_2} = \frac{x_3}{c_3t_3} = \ldots\ldots$$

where $x_1, x_2, x_3, \ldots\ldots$ are the interests earned on the capitals $c_1, c_2, c_3, \ldots\ldots$ in $t_1, t_2, t_3, \ldots\ldots$ months respectively.

Therefore,

$$\frac{x_1}{c_1t_1} = \frac{x_2}{c_2t_2} = \frac{x_3}{c_3t_3} = \ldots = \frac{x_1+x_2+x_3+\ldots}{c_1t_1+c_2t_2+c_3t_3+\ldots}$$

$$= \frac{m}{c_1t_1+c_2t_2+c_3t_3+\ldots}$$

or $$x_1 = \frac{m\, c_1t_1}{c_1t_1+c_2t_2+c_3t_3+\ldots}, \quad \text{etc.}$$

This formula is given by Mahâvîra for the solution

[1] The Sanskrit term used is *phala* (fruit).

[2] The Sanskrit term used is *guṇa* (multiple).

[3] *GSS*, p. 70(38).

of the above problem.[1]

(2) "(Sums represented by) 10, 6, 3 and 15 are the (various given) amounts of interest, and 5, 4, 3 and 6 are the (corresponding) months (for which the interests have accrued); the sum of the (corresponding) capital amounts is seen to be 140. (Assuming the rate of interest to be the same in each case, find out these capital amounts)."[2]

(3) "Here (in this problem) the (given) capital amounts are 40, 30, 20 and 50; and 10, 6, 3 and 15 are the (corresponding) amounts of interest; 18 is the quantity representing the mixed sum of the respective periods of time. (Find out these periods separately, assuming the rate of interest to be the same in each case)."[3]

(4) "The interest on 80 for 3 months is unknown; $7\frac{4}{5}$ is the mixed sum of that (unknown quantity taken as the) capital lent out and of the interest thereon for 1 year. What is the capital here and what the interest?"[4]

(5) "The mixed sums (capital+interest) are 50, 58 and 66, and the months (during which interests have accrued) are 5, 7 and 9 (respectively). Find out what

[1] *GSS*, p. 70(37). The formula clearly shows that Mahâvîra knew the algebraic identity

$$\frac{a}{b}=\frac{c}{d}=\frac{e}{f}=\ldots=\frac{a+c+e+\ldots}{b+d+f+\ldots}.$$

[2] *GSS*, p. 70(40). The solution is given by Rule 39 on the same page.

[3] *GSS*, p. 70(43). The solution is given by Rule 42 on the same page.

[4] *GSS*, p. 71(46). This is similar to Âryabhaṭa's problem given before (p. 217).

the interest is (in each case, the capital being the same)?"[1]

(6) "The mixed sums of the capital and periods of interest are 21, 23, and 25; here (in this problem) the amounts of interest are 6, 10 and 14. What is the common capital?"[2]

(7) "Borrowing at the rate of 6 per cent and then lending out at the rate of 9 per cent, one obtains in the way of differential gain 81 at the end of 3 months. What is the capital (utilised here)?"[3]

(8) "The monthly interest on 60 is exactly 5. The capital lent out is 35; the (amount of the) instalment (to be paid) is 15 in (every) 3 months. What is the time of discharge of that debt?"[4]

(9) "The mixed sum (of the capital amounts lent out) at the rates of 2, 6 and 4 per cent per mensem is 4400. Here the capital amounts are such as have equal amounts of interest accruing after 2 months. What (are the capital amounts lent, and what is the equal interest)?"[5]

(10) "A certain person gives once in 12 days an instalment of $2\frac{3}{5}$, the rate of interest being 3 per cent (per mensem). What is the capital amount of the debt discharged in 10 months?"[6]

(11) "The total capital represented by 8520 is invested (in parts) at the (respective) rates of 3, 5 and 8 per cent (per month). Then, in this investment, in 5

[1] *GSS*, p. 71 (48). The solution requires the use of the identity

$$\frac{a}{b} = \frac{c}{d} = \frac{a-c}{b-d}.$$

[2] *GSS*, p. 72(52).

[3] *GSS*, p. 72(55).

[4] *GSS*, p. 73(59).

[5] *GSS*, p. 73(61).

[6] *GSS*, p. 73(65).

months the capital amounts lent out are, on being diminished by the respective amounts of interest, (found to be) equal in value. (What are the respective amounts invested thus?)"[1]

(12) "The total capital represented by 13740 is invested (in parts) at the (respective) rates of 2, 5 and 9 per cent (per month), then, in this investment, in 4 months the capital amounts lent out are, on being combined with the (respective) amounts of interest, (found to be) equal in value. (What are the respective amounts thus invested?)"[2]

(13) "A certain man borrows a certain (unknown) sum of money at an interest of 5 per cent per month. He pays the debt in instalments, due every $\frac{3}{5}$ of a month. The instalments begin with 7 and increase in arithmetical progression, with 7 as the common-difference. 60 is the maximum amount of instalment. He gives in the discharge of his debt the sum of a series in arithmetical progression consisting of $\frac{60}{7}$ terms. After the payment of each instalment, interest is charged only on that part of the principal which remains to be paid. What is the total payment corresponding to the sum of the series, what is the interest (which he paid), what is the time of the discharge of the debt, (and what is the principal sum borrowed)?"[3]

Barter and Exchange. The Hindu name for barter is *bhâṇḍa-prati-bhâṇḍa* ("commodity for commodity"). All the Hindu works on *pâṭîgaṇita* contain problems relating to the exchange of commodities. It is pointed out in these works that problems on barter are cases of compound proportion, and can be solved by the

[1] *GSS*, p. 74(67).
[2] *GSS*, p. 74(67).
[3] *GSS*, pp. 74f, (72-73½). The text of the problem is very obscure. The translation given here is after Rangacarya.

Rule of Five, etc. A typical problem on barter is the following:

"If three hundred mangoes be had in this market for one *dramma*, and thirty ripe pomegranates for a *paṇa*, say quickly, friend, how many (pomegranates) should be had in exchange for ten mangoes?"[1]

Other Types of Commercial Problems. Of various other types of commercial problems found in the Hindu works may be mentioned (1) problems on partnership and proportionate division, and (2) problems relating to the calculation of the fineness of gold.[2] Most of these problems are essentially of an algebraic character, but they are included in *pâṭîgaṇita* (arithmetic). The formulæ giving the solution of each type of examples precede the examples. These formulæ are too numerous to be mentioned. The following examples, however, will illustrate the nature and the scope of such problems:

(1) A horse was purchased by (nine) dealers in partnership, whose contributions were one, etc., upto nine; and was sold by them for five less than five hundred. Tell me what was each man's share of the sale-proceed.

(2) Four colleges, containing an equal number of pupils, were invited to partake of a sacrificial feast. A fifth, a half, a third and a quarter (of the total number of pupils in the college) came from the respective colleges to the feast; and added to one, two, three and four, they were found to amount to eighty-seven; or, with those deducted, they were sixty-seven. Find the actual number of the pupils that came from each college.

[1] *L*, p. 20.

[2] Such problems are found in the *Lîlâvatî*, the *Gaṇita-sâra-saṅgraha*, the *Triśatikâ*, etc.

(3) Three (unequal) jars of liquid butter, of water and of honey, contained thirty-two, sixty and twenty-four *pala* respectively: the whole was mixed together and the jars filled again. Tell me the quantity of butter, of water and of honey in each jar.[1]

(4) According to an agreement three merchants carried out the operations of buying and selling. The capital of the first consisted of six *purâṇa*, that of the second of eight *purâṇa*, but that of the third was unknown. The profit obtained by these men was 96 *purâṇa*. In fact the profit obtained by him (the third person) on his unknown capital happened to be 40 *purâṇa*. What was the amount thrown by him into the transaction and what was the profit of each of the other two merchants?[2]

(5) There were four merchants. Each of them obtained from the others half of what he had with him (at the time of the respective transfers of money). Then they all became possessed of equal amounts of money. What was the measure of money each had to start with?[3]

(6) A great man possessing powers of magical charm and medicine saw a cock fight going on, and spoke separately in confidential language to both the owners of the cocks. He said to one, "If your bird wins, then you give the stake-money to me. If, however, your bird loses then I shall give you two-thirds of that stake-money." He went to the owner of the other cock and promised to give three-fourths (of his stake-money on similar conditions). In each case the gain to him could be only 12 (gold-pieces). Tell me, O ornament

[1] This and the two previous examples are given by Pṛthudakasvâmî to illustrate Rule 16 of the *gaṇitâdhyâya* of the *Brâhma-sphuṭa-siddhânta*.

[2] *GSS*, p. 94(223-5).

[3] *GSS*, p. 99(267½).

on the head of mathematicians, the money each of the cock-owners had staked.[1]

(7) The mixed price of 9 citrons and 7 fragrant wood-apples is 107; again the mixed price of 7 citrons and 9 fragrant wood-apples is 101. O arithmetician, tell me quickly the price of a citron and of a wood-apple, having distinctly separated those prices.[2]

(8) Pigeons are sold at the rate of 5 for 3 (*paṇa*), *sârasa* birds at the rate of 7 for 5 (*paṇa*), swans at the rate of 9 for 7 (*paṇa*) and peacocks at the rate of 3 for 9 (*paṇa*). A certain man was told to bring at these rates 100 birds for 100 (*paṇa*) for the amusement of the king's son, and was sent to do so. What (amount) does he give for each (of the various kinds of birds that he buys)?[3]

(9) There are 1 part (of gold) of 1 *varṇa*, 1 part of 2 *varṇa*, 1 part of 3 *varṇa*, 2 parts of 4 *varṇa*, 4 parts of 5 *varṇa*, 7 parts of 14 *varṇa*, and 8 parts of 15 *varṇa*. Throwing these into the fire, make them all into one (mass), and then (say) what the *varṇa* of the mixed gold is. This mixed gold is distributed among the owners of the foregoing parts. What does each of them get?[4]

(10) Three pieces of gold, of 3 each in weight, and of 2, 3 and 4 *varṇa* (respectively), are added to (an unknown weight of) gold of 13 *varṇa*. The resulting *varṇa* comes to be 10. Tell me, O friend, the measure (of the unknown weight) of gold.[5]

[1] *GSS*, pp. 99-100($270\frac{1}{2}$-$2\frac{1}{2}$).
[2] *GSS*, p. 84($140\frac{1}{2}$-$2\frac{1}{2}$).
[3] *GSS*, p. 85(152-3).
[4] *GSS*, p. 88 (170-$1\frac{1}{2}$).
[5] *GSS*, p. 89(181). Similar examples occur in the *Triśatikâ* (p. 26) and the *Lîlâvatî* (p. 25).

14. MISCELLANEOUS PROBLEMS

Regula Falsi. The rule of false position is found in all the Hindu works.[1] Bhâskara II gives prominence to the method and calls it *iṣṭa-karma* ("rule of supposition"). He describes the method thus:

"Any number, assumed at pleasure, is treated as specified in the particular question, being multiplied and divided, increased or diminished by fractions (of itself); then the given quantity, being multiplied by the assumed number and divided by that (which has been found) yields the number sought. This is called the process of supposition."[2]

Śrîdhara takes the assumed number to be one.[3] Mahâvîra gives a large variety of problems to which he applies the rule.[4] Gaṇeśa in his commentary on the *Lîlâvatî* remarks, "In this method, multiplication, division, and fractions only are employed." The following examples will illustrate the nature of the problems solved by the rule of supposition:

(1) Out of a heap of pure lotus flowers, a third, a fifth, a sixth were offered respectively to the gods Śiva, Viṣṇu and Sûrya and a quarter was presented to Bhavânî. The remaining six were given to the venerable preceptor. Tell quickly the number of lotuses.[5]

(2) The third part of a necklace of pearls, broken in

[1] The method originated in India and went to Europe through Arabia. There is a mediæval MS., published by Libri in his *Histoire*, I, 304 and possibly due to Rabbi ben Ezra in which the method is attributed to the Hindus. For further details and references, see Smith, *History*, II, p. 437, foot-note 1.

[2] *L*, p. 10.

[3] See the rule on *stambhoddeśa*, *Triś*, p. 13.

[4] These problems occur in chapters iii and iv of the *Gaṇita-sâra-saṅgraha*.

[5] *L*, p. 11. *Cf. GSS*, p. 48 (7).

an amorous struggle, fell to the ground; its fifth part rested on the couch; the sixth part was saved by the wench; and the tenth part was taken by her lover: six pearls remained strung. Say, of how many pearls was the necklace composed?[1]

(3) One-twelfth part of a pillar, as multiplied by $\frac{1}{30}$ part thereof, was to be found under water; $\frac{1}{20}$ of the remainder, as multiplied by $\frac{3}{16}$ thereof, was found buried in the mire below; and 20 *hasta* of the pillar were found in the air (above the water). O friend, give out the length of the pillar.[2]

(4) A number of parrots descended on a paddy field, beautiful with crops bent down through the weight of ripe corn. Being scared away by men, all of them suddenly flew off. One-half of them went to the east, one-sixth went to the south-east; the difference between those that went to the east and those that went to the south-east, diminished by half of itself and again diminished by half of this (resulting difference), went to the south. The difference between those that went to the south and those that went to the south-east diminished by two-fifths of itself went to the south-west; the difference between those that went to the south and those that went to the south-west, went to the west; the difference between those that went to the south-west and those that went to the west, together with three-sevenths of itself went to the north-west; the difference between those that went to the north-west and those that went to the west together with seven-eighths of itself, went to the north; the sum of those that went to the north-west and those that went to the north, diminished by two-thirds of itself went to the north-east; and 280 parrots were found to

[1] *Triś*, p. 14, *cf. GSS*, p. 49 (17-22) for a similar example.
[2] *GSS*, p. 55(60). *Cf. Triś*, p. 13.

remain in the sky above. How many were the parrots in all?[1]

The Method of Inversion. The method of inversion called *vilomagati* ("working backwards") is found to have been commonly used in India from very early times. Thus Āryabhaṭa I says:

"In the method of inversion multipliers become divisors and divisors become multipliers, addition becomes subtraction and subtraction becomes addition."[2]

Brahmagupta's description is more complete. He says:

"Beginning from the end, make the multiplier divisior, the divisor multiplier; (make) addition subtraction and subtraction addition; (make) square square-root, and square-root square; this gives the required quantity."[3]

The following examples will illustrate the nature of problems solved by the above method:

(1) What is that quantity which when divided by 7, (then) multiplied by 3, (then) squared, (then) increased by 5, (then) divided by $\frac{3}{5}$, (then) halved, and then reduced to its square-root happens to be the number 5?[4]

(2) The residue of degrees of the sun less three, being divided by seven, and the square-root of the quotient extracted, and the root less eight multiplied by nine, and to the product one being added, the amount is

[1] *GSS*, pp. 48f (12-16).

[2] *Ā, Gaṇitapâda*, 28.

[3] *BrSpSi*, p. 301. The method occurs also in *GSS*, p. 102 (286); *MSi*, p. 149; *L*, p. 9; etc.

[4] *GSS*, p. 102 (287). Examples of this type are very common in Hindu arithmetic. They were also very common in Europe. Smith in his *History*, II, quotes two such problems from an American arithmetic of the 16th century.

a hundred. When does this take place on a Wednesday?[1]

Problems on Mixture. The Hindu works on *pâṭîgaṇita* contain a chapter relating to problems on mixture (*miśraka-vyavahâra*). Miscellaneous problems on interest, problems on allegation, and various other types of problems, in which quantities are to be separated from their mixture, form the subject matter of *miśraka-vyavahâra*. A chapter "on mixture" (*De' mescolo*) is found in early Italian works on arithmetic, evidently under Hindu influence.[2]

Some of the problems of this chapter are determinate and some are indeterminate. A few relating to interest and allegation have already been given.[3] The following are some others:

(1) In the interior of a forest, 3 heaps of pomegranates were divided (equally) among 7 travellers, leaving 1 fruit as remainder; 7 (of such heaps) were divided among 9, leaving a remainder of 3 (fruits), again 5 (of such heaps) were divided among 8, leaving 2 fruits as remainder. O mathematician, what is the numerical value of a heap?[4]

(2) On a certain man bringing mango fruits home, his elder son took one fruit first and then half of what remained. The younger son did similarly with what was left. He further took half of what was left thereafter; and the other took the other half. Find the number of fruits brought by the father?[5]

[1] Colebrooke, cha, p. 333 (18).

[2] Smith, *History*, II, p. 588, note 4.

[3] See commercial problems, pp. 216ff; also problems on proportionate division (*prakṣepa-karaṇa*): *Triś*, p. 26; *GSS*, p. 75(79½); *MSi*, pp. 154-155.

[4] *GSS*, p. 82 (128½). Such problems are given under the rule of *vallikâ-kuṭṭîkâra* by Mahâvîra.

[5] *GSS*, p. 82 (131½).

(3) A certain lay follower of Jainism went to a Jina temple with four gate-ways, and having taken (with him) fragrant flowers offered them in worship with devotion (at each gate). The flowers in his hand were doubled, trebled, quadrupled and quintupled (respectively in order) as he arrived at the gates (one after another). The number of flowers offered by him was sixty[1] at each gate. How many flowers were originally taken by him?

(4) The first man has 16 azure-blue gems, the second has 10 emeralds, and the third has 8 diamonds. Each among them gives to each of the others 2 gems of the kind owned by himself; and then all three men come to be possessed of equal wealth. What are the prices of those azure-blue gems, emeralds and diamonds?[2]

(5) In what time will four fountains, being let loose together, fill a cistern, which they would severally fill in a day, in half a day, in a quarter and in a fifth part of a day?[3]

Problems involving Solution of Quadratic Equations. The solution of the quadratic equation has been known in India from the time of Āryabhaṭa I (499). Problems on interest requiring the solution of the quadratic equation have already been mentioned. Mahâvîra and Bhâskara II give many other problems. Mahâvîra divides these problems into two classes: (*i*) those that involve square-roots (*mûla*) and (*ii*) those

[1] *GSS*, p. 79 (112½-113½). The printed text has *pañca* ("five"). According to it the answer is 43/12 which appears absurd. There are some other problems in the printed edition which give such absurd results. All those are, we presume, due to the defects of the mss. consulted by the editor. So here we have made the emendation 'sixty.'

[2] *GSS*, p. 87 (165-166).

[3] *BrSpSi*, p. 177 (com.); *L*, p. 23.

that involve the square (*varga*) of the unknown. The first type gives a single positive answer, while the second type has two answers corresponding to the two roots of the quadratic. Bhâskara II deals with the first type of problems only in his *pâṭîgaṇita*, the *Lîlâvatî*. The second type of problems, involving the square of the unknown has been treated by him in his *Bîjagaṇita* (algebra). The following examples will illustrate the nature and scope of such problems:

Problems involving the square-root:

(1) One-fourth of a herd of camels was seen in the forest; twice the square-root of that had gone to mountain slopes; and three times five camels were found to remain on the bank of a river. What was the numerical measure of that herd of camels?[1]

(2) Five and one-fourth times the square-root (of a herd) of elephants are sporting on a mountain slope; five-ninths of the remainder sport on the top of the mountain; five times the square-root of the remainder sport in a forest of lotuses; and there are six elephants then (left) on the bank of a river. How many are the elephants?[2]

(3) In a garden beautified by groves of various kinds of trees, in a place free from all living animals, many

[1] *GSS*, p. 51 (34). The problem belongs to the type of the *mûla-jâti*, and leads to an equation of the form $x-(bx+c\sqrt{x}+a)=0$. The method of solution is given in *GSS*, p. 50 (33).

[2] *GSS*, p. 52 (46). The problem is of the *śeṣa-mûla* variety. It gives the equation

$$x-\frac{21}{4}\sqrt{x}-\frac{5}{9}\left(x-\frac{21}{4}\sqrt{x}\right)-5\sqrt{x-\frac{21}{4}\sqrt{x}-\frac{5}{9}\left(x-\frac{21}{4}\sqrt{x}\right)}=6.$$

Mahâvîra reduces it by putting $z=x-\frac{21}{4}\sqrt{x}-\frac{5}{9}\left(x-\frac{21}{4}\sqrt{x}\right)$. to $z-5\sqrt{z}=6$. In the general case a similar equation is again obtained, which is again reduced, and so on till the equation is reduced to the form, $x-b\sqrt{x}=d$, from which x can be easily obtained.

ascetics were seated. Of them the number equivalent to the square-root of the whole collection were practising *yoga* at the foot of a tree. One-tenth of the remainder, the square-root (of what remained after this), $\frac{1}{9}$ (of what remained after this), then the square-root (of what remained after this), $\frac{1}{8}$ (of what remained after this), the square-root (of what remained after this), $\frac{1}{7}$ (of what remained after this), the square-root (of what remained after this), $\frac{1}{6}$ (of what remained after this), the square-root (of what remained after this), $\frac{1}{5}$ (of what remained after this), the square-root (of what remained after this)—these parts consisted of those who were learned in the teaching of literature, in religious law, in logic, and in politics, as also of those who were versed in controversy, prosody, astronomy, magic, rhetoric and grammar, as well as of those who possessed an intelligent knowledge of the twelve varieties of the *aṅga-śâstra*; and at last 12 ascetics were seen (to remain without being included among those mentioned before). O excellent ascetic, of what numerical value was this collection of ascetics?[1]

(4) A single bee (out of a swarm of bees) was seen in the sky; $\frac{1}{5}$ of the remainder (of the swarm), and $\frac{1}{4}$ of the remainder (left thereafter) and again $\frac{1}{3}$ of the remainder (left thereafter) and a number of bees equal to the square-root of the numerical value of the swarm, were seen in lotuses; and two bees were on a mango tree. How many were there?[2]

(5) Four times the square-root of half the number of a collection of boars went to a forest wherein tigers

[1] *GSS*, p. 52 (42-45). The problem is of the same variety as the above one. The substitution will have to be made 6 times to reduce the resulting equation.

[2] *GSS*, p. 53 (48). This problem is of the *dviragra-śeṣa-mûla* variety.

were at play; 8 times the square-root of $\frac{1}{10}$ of the remainder went to a mountain; and 9 times the square-root of $\frac{1}{2}$ of the (next) remainder went to the bank of a river; and boars equivalent in (numerical) measure to 56 were seen to remain in the forest. Give the numerical measure of all those boars.[1]

(6) The sum of two (quantities, which are respectively equivalent to the) square-root (of the numerical value) of a collection of swans and (the square-root of the same collection) as combined with 68, amounts to 34. How many swans there are in that collection?[2]

(7) Pârtha (Arjuna), irritated in fight, shot a quiver of arrows to slay Karṇa. With half his arrows, he parried those of his antagonist, with four times the square-root of the quiver-full, he killed his horses; with three he demolished the umbrella, standard and bow; and with one he cut off the head of his foe. How many were the arrows, which Arjuna let fly?[3]

(8) The square-root of half the number of a swarm of bees is gone to a shrub of jasmin; and so are eight-ninths of the whole swarm; a female is buzzing to one remaining male that is humming within a lotus, in which he is confined, having been allured to it by its

[1] *GSS*, p. 54 (56). This problem is of the *aṁśa-mûla* variety, wherein fractional parts of square-roots are involved. The problems give equations of the form

$$x-a_1\sqrt{b_1x} - a_2\sqrt{b_2(x-a_1\sqrt{b_1x})}$$

$$-a_3\sqrt{b_3[(x-a_1\sqrt{b_1x}) - a_2\sqrt{b_2(x-a_1\sqrt{b_1x})}} - \ldots = k.$$

By repeated substitutions Mahâvîra reduces the equation to the form $x - A\sqrt{Bx} - c = 0$.

[2] *GSS*, p. 56 (68). This problem is of the *mûla-miśra* variety, wherein the sum of square-roots is involved. It gives an equation of the form $\sqrt{x} + \sqrt{x \pm d} = m$.

[3] *L*, p. 16.

fragrance at night. Say, lovely woman, what is the number of bees.[1]

Problems involving the square of the unknown:

(9) One-twelfth part of a pillar, as multiplied by $\frac{1}{30}$ part thereof, was found under water; $\frac{1}{20}$ of the remainder, as multiplied by $\frac{3}{16}$ thereof, was found buried in the mire, and 20 *hasta* of the pillar were found in the air. O friend, give the measure of the length of the pillar.[2]

(10) A number of elephants (equivalent to) $\frac{1}{10}$ of the herd minus 2, as multiplied by the same ($\frac{1}{10}$ of the herd minus 2), is found playing in a forest of *sallakî* trees. The remaining elephants of the herd equal in number to the square of 6 are moving on a mountain. How many are the elephants?[3]

15. THE MATHEMATICS OF ZERO

It has been shown that the zero was invented in India about the beginning of the Christian era to help the writing of numbers in the decimal scale. The Hindu mind did not rest satisfied till it evolved the complete arithmetic of zero. The Hindus included zero among the numbers (*saṅkhyâ*), and it was used

[1] *L*, p. 16.

[2] *GSS*, p. 55 (60). The problem gives the equation

$$\left(x - \frac{x^2}{12.30}\right) - \frac{1.3}{20.16}\left(x - \frac{x^2}{12.30}\right)^2 = 20.$$

Also solved by *regula falsi*. Mahâvîra puts $\left(x - \frac{1}{12.30}x^2\right) = z$, and then solves the quadratic

$$z - \frac{3}{320}z^2 = 20.$$

The roots of this are then used to get the values of x.

[3] *GSS*, p. 55 (63).

in their arithmetic at the time when the original of the Bakhshâlî Manuscript was written, about the third century A.D. The operation of addition and subtraction of zero are incidentally mentioned in the *Pañca-siddhântikâ* of Varâhamihira (505). The complete decimal arithmetic is found in the commentary of Bhâskara I (*c.* 525) on the *Âryabhaṭîya*. The results of operations by zero are found stated in the work of Brahmagupta (628) and in all later mathematical treatises. The treatment of zero in the arithmetic of the Hindus is different from that found in their algebra. In order, therefore, to bring out this difference clearly, we give separately the results found in *pâṭîgaṇita* (arithmetic) and in *bîjagaṇita* (algebra).

Zero in Arithmetic. The Hindus in their arithmetic define zero as the result of the operation

$$a - a = 0$$

This definition is found in Brahmagupta's work[1] and is repeated in all later works. It is directly used in the operation of subtraction. In carrying out arithmetical operations, the results of the operations of addition, subtraction and multiplication of zero and by zero are required. The Hindus did not recognise the operation of division by zero as valid in arithmetic; but the division of zero by a number was recognised as valid. Nârâyaṇa in his *pâṭîgaṇita* (arithmetic) has clearly stated this distinction:

"Here in *pâṭîgaṇita*, division by zero is not recognised, and therefore, it is not mentioned here. As it is of use in *bîjagaṇita* (algebra), so I have mentioned division by zero in my *Bîjagaṇita*."[2]

[1] *BrSpSi*, p. 309. *Cf.* B. Datta, *BCMS*, XVIII, pp. 165-176 for some other details regarding operations with zero.

[2] *GK*, remark subjoined to i 30.

The following are the results of the operations in relation to zero mentioned in Hindu works on *pâṭî-gaṇita*:

Śrîdhara says:

"In addition cipher makes the sum equal to the additive; when cipher is subtracted (from a number), there is no change in the number. In multiplication and other[1] operations on zero the result is zero. Multiplication of a number by cipher also gives zero."[2]

Âryabhaṭa II, in the chapter on *pâṭîgaṇita* in his *Mahâ-siddhânta*, writes:

"If zero is added to a number, the number is unchanged; the same is true when zero is subtracted. In the operation of multiplication or division of zero (by another number) or in evolution and involution (of zero), the result is zero."[3]

Nârâyaṇa in his *pâṭîgaṇita* states:

"When zero is added to or subtracted from a number, it (the number) remains unchanged; that (number) multiplied by zero becomes zero. In the multiplication and other operations on zero, the result is zero. When a quantity is added to zero the result is that which is added."[4]

Mahâvîra in his *Gaṇita-sâra-saṁgraha* writes:

"A number multiplied by zero is zero, and that number remains unchanged when it is divided by, combined with, or diminished by zero. Multiplication

[1] By other operations are meant: (1) division of zero by a number, (2) squaring and square-root of zero, (3) cubing and cube-root of zero—in the sequence in which these operations are mentioned in Hindu arithmetic.

[2] *Triś*, p. 4. It should be noted that distinction is made between $a \times 0$ and $0 \times a$, although the result in each case is zero.

[3] *MSi*, p. 146.

[4] *GK*, i. 30.

and other operations on zero give rise to zero and in the operation of addition, zero becomes the same as what is added to it."[1]

Mahâvîra's statement that a number remains unchanged when divided by zero is obviously incorrect. The correct result of the operation was known to Brahmagupta who preceded him by more than two centuries. It is, therefore, strange to find Mahâvîra making such a statement. He probably thought that, so far as arithmetic was concerned, division by zero was no division at all.

Zero in Algebra. The earliest treatment of zero in algebra is found in the *Brâhma-sphuṭa-siddhânta*[2] (628):

"Negative less cipher is negative; positive (less cipher is) positive; cipher (less cipher is) nought....

"The product of cipher and negative, or of cipher and affirmative, is cipher; of two ciphers is cipher.

"Cipher, divided by cipher is nought.... Positive or negative divided by cipher is *taccheda* (a fraction with that for denominator), or cipher divided by negative or affirmative (is either zero or is expressed by a fraction with zero as numerator and the finite quantity as denominator)."[3]

Bhâskara II in the *Lîlâvatî* as well as in his *Bîja-gaṇita* gives the result of the operations with zero. In the *Lîlâvatî* he writes:

"In addition cipher makes the sum equal to the additive. In involution and evolution the result is

[1] *GSS*, p. 6. (49).

[2] *BrSpSi*, pp. 309-310, rules 31-35. These rules contain also, the rules of the operations with the positive and the negative in algebra. The results relating to zero only are mentioned here.

[3] The result of the operation given in the bracket is according to the interpretation of S. Dvivedî.

cipher. A number divided by zero is *kha-hara* (that number with zero as denominator). The product of (a number and) zero is zero, but it must be retained as a multiple of zero (*kha-guṇa*), if any further operations impend. Zero having become a multiplier (of a number), should zero afterwards become a divisor, the number must be understood to be unchanged. So likewise any number, to which zero is added, or from which it is subtracted (is unaltered)."[1]

In the *Bījagaṇita*, the same results are given with the addition that if a quantity is subtracted from zero, its sign is reversed, while in the case of addition the sign remains the same.

Zero as an Infinitesimal. It will be observed that Brahmagupta directs that the results of the operations $x \div 0$ and $0 \div x$ should be written as $\frac{x}{0}$ and $\frac{0}{x}$ respectively. It is not possible to tell exactly what he actually meant by these forms. It seems that he did not specify the actual value of these forms, because the value of the variable x is not known. Moreover, the zero seems to have been considered by him as an infinitesimal quantity which ultimately reduces to nought. If this surmise be correct, Brahmagupta is quite justified in stating the results as he has done.

The idea of zero as an infinitesimal is more in evidence in the works of Bhâskara II. He says: "The product of (a number and) zero is zero, but the number must be retained as a multiple of zero (*kha-guṇa*), if any further operations impend." He further remarks that this operation is of great use in astronomical calculations. It will be shown in the section on Calculus, that Bhâskara II has actually used quantities which ultimately tend to zero, and has successfully evaluated the differential coefficients of certain functions. He has, moreover,

[1] *L*, p. 8.

used the infinitesimal increment $f'(x)\delta x$ of the function $f(x)$, due to a change δx in x.

The commentator Kṛṣṇa proves the result $0 \times a = 0 = a \times 0$ as follows:

"The more the multiplicand is diminished, the smaller is the product; and, if it be reduced in the utmost degree, the product is so likewise: now the utmost diminution of a quantity is the same with the reduction of it to nothing; therefore, if the multiplicand be nought, the product is cipher. In like manner, as the multiplier decreases, so does the product; and, if the multiplier be nought, the product is so too."

In the above zero is conceived of as the limit of a diminishing quantity.

Infinity. The quotient of division by zero of a finite quantity has been called by Bhâskara II as *kha-hara*, which is synonymous with *kha-cheda* (the quantity with zero as denominator) of Brahmagupta. Regarding the value of the *kha-hara*, Bhâskara II remarks:

"In this quantity consisting of that which has cipher for its divisor, there is no alteration, though many may be inserted or extracted; as no change takes place in the infinite and immutable God, at the period of the destruction or creation of worlds, though numerous orders of beings are absorbed or put forth."[1]

From the above it is evident that Bhâskara II knew that $\frac{a}{0} = \infty$ and $\infty + k = \infty$.

[1] *BBî*, pp. 5-6. G. Thibaut (*Astronomie, Astrology und Mathematik*,, Strasbourg, 1899, p. 72) thought that this passage was an interpolation. There appears no justification for considering this as an interpolation, as the passage occurs in the oldest known commentary and in all copies of the work so far found. *Cf.* Datta, *l.c.*, p. 174.

Gaṇeśa remarks that $\frac{a}{0}$ is "an indefinite and unlimited or infinite quantity: since it cannot be determined how great it is. It is unaltered by the addition or subtraction of finite quantities: since in the preliminary operation of reducing both fractional expressions to a common denominator, preparatory to taking their sum or difference, both numerator and denominator of the finite quantity vanish."

Kṛṣṇa remarks:

"As much as the divisor is diminished, so much is the quotient increased. If the divisor is reduced to the utmost, the quotient is to the utmost increased. But, if it can be specified, that the amount of the quotient is so much, it has not been raised to the utmost: for a quantity greater than that can be assigned. The quotient, therefore, is indefinitely great, and is rightly termed infinite."

Regarding the proof of $\frac{a}{0} \pm k = \frac{a}{0}$ Kṛṣṇa makes the same remarks as Gaṇeśa. He, however, goes a step further when he says that

$$\frac{a}{0} = \frac{b}{0}.$$

This is illustrated by him through the instance of the shadow of a gnomon, which at sun-rise and sun-set is infinite; and is equally so whatever height be given to the gnomon, and whatever number be taken for the radius. "... Thus, if the radius be 120; and the gnomon be 1, 2, 3 or 4; the expression deduced from the proportion, as sine of sun's altitude is to sine of zenith distance, so is gnomon to shadow, becomes $\frac{120}{0}$, $\frac{240}{0}$, $\frac{360}{0}$ or $\frac{480}{0}$. Or, if the gnomon be, as it is usually framed, 12 fingers, and radius be taken as 3438, 120, 100 or 90, the expression will be $\frac{41256}{0}$, $\frac{1440}{0}$, $\frac{1200}{0}$ or $\frac{1080}{0}$, which are all alike infinite."[1]

Indeterminate Forms. Brahmagupta has made the incorrect statement that

$$\frac{0}{0}=0$$

Bhâskara II has sought to correct this mistake of Brahmagupta. According to him

$$\lim_{\varepsilon \to 0} \frac{a.\varepsilon}{\varepsilon} = a.$$

His language, however, in stating this result is defective, for he calls the infinitesimal ε zero, not being in possession of a suitable technical term. That, in the above case, he actually meant by zero a small quantity tending to the limiting value zero, is abundantly clear from the use he makes of the result in his Astronomy. Taylor[2] and Bapu Deva Ŝâstrî[3] are also of this opinion.

Bhâskara has given three illustrative examples. They are:

(*i*) Evaluate $\dfrac{\left(x\times 0+\dfrac{x\times 0}{2}\right)}{0}=63.$

From this he derives the result $x=14$, which is correct if we consider $0=\varepsilon$, a small quantity tending to zero. His other examples are:

(*ii*) $\left\{\left(\frac{x}{0}+x-9\right)^2+\left(\frac{x}{0}+x-9\right)\right\}0=90$

giving $x=9$; and

[1] All the above passages are taken from the respective commentaries. They have been noted by Colebrooke, *l.c.*

[2] *Lîlâwatî*, Bombay, 1816, p. 29.

[3] His *Bîja-gaṇita* (in Hindi), Pt. I, Benares, 1875, p. 179 *et sq.*

$$(iii)\ \{(x+\tfrac{x}{2})\times 0\}^2+2\{(x+\tfrac{x}{2})\times 0)\}\div 0 = 15,$$

giving $x=2$.[1]

Bhâskara II's result

$$\frac{a}{0}\times 0 = a$$

is, however, not quite correct, as the form is truly indeterminate and may not always have the value a. His attempt, however, at such an early date to assign a meaning to the form $\frac{0}{0}$, and his partial solution of the problem are very creditable, seeing that in Europe mathematicians made similar mistakes upto the middle of the nineteenth century A. D.[2]

[1] The answers of this and the previous example are incorrect because 0^2 has been taken to be equal to 0.

[2] Martin Ohm (1828) says: "If a is not zero, but b is zero, then the quotient a/b has no meaning" for the quotient "multiplied by zero gives only zero and not a, as long as a is not zero." *Lehrbuch der niedern Analysis*, Vol. I, Berlin, 1828, pp. 110, 112.

BIBLIOGRAPHY

OF SANSKRIT MATHEMATICAL WORKS

1. *Âpastamba Śulba* Sûtra *by* Âpastamba (*c.* 400 B.C.). Edited with the commentary of Kapardisvâmî, Karavindasvâmî and of Sundararâja *by* D. Srinivasachar and V.S. Narasimhachar, University of Mysore Sanskrit Series, 1931; *by* A. Bürk, with German translation and note and comments in *Zeitschrift der deutschen morgenländischen Gesselschaft*, LV, 1902, pp. 543-591; LVI, 1903, pp. 327-391.

2. *Ârṣa Jyotiṣa.* Edited *by* Sudhakara Dvivedi with his own commentary, Benares, 1906.

3. *Âryabhaṭîya by* Âryabhaṭa I (499). Edited with the commentary, entitled *Bhatadîpikâ*, of Parameśvara (1430) *by* H. Kern (Leiden, 1875); *by* Udai Narayan Singh, with explanatory notes in Hindi (Muzaffarpur, 1906); with the commentary, entitled *Mahâbhâṣya*, of Nîlakaṇṭha (1500) on the last three chapters *by* K. Sambasiva Sastri (Trivandrum, Part I, 1930; Part II, 1931; Part III, *in press*). Translations: *by* P. C. Sengupta ("The Âryabhaṭîyam," in the *Journal of the Department of Letters in the University of Calcutta*, XVI, 1927); *by* W. E. Clark (*The Âryabhaṭîya of Âryabhaṭa*, Chicago, 1930). The second chapter of the book has been translated also *by* L. Rodet ("Leçons de calcul d'Âryabhaṭa," in *Journal Asiatique*, XIII (7), 1878; reprint, Paris, 1879) and *by* G. R. Kaye ("Notes on Indian Mathematics, No. 2—Âryabhaṭa," in *Journal of the Asiatic Society of Bengal*, IV, 1908). Other commentaries in MSS: (i) *by* Bhâskara I (522); (ii) *Bhataprakâśikâ by* Sûryadeva Yajvâ (12th century).

4. *Atharvan Jyotiṣa.* Edited *by* Bhagvad Datta, Lahore, 1924.

5. *Bakhshâlî Manuscript—A Study in Mediæval Mathematics*, Parts I, II and III edited *by* G. R. Kaye, Calcutta, 1927, 1933.

6. *Baudhâyana Śulba* Sûtra *by* Baudhâyana (*c.* 800 B.C.). Edited *by* G. Thibaut, with English translation, critical notes and extracts from the commentary of Dvârakânâtha Yajvâ, in

the *Paṇḍit* (Old Series, IX and X, 1874-5; New Series, I, 1877). The text appears in the edition of the *Baudhâyana Śrauta Sûtra* (as its 30th chapter) *by* W. Caland (in 3 volumes, Calcutta, 1904, 1907, 1913).

7. *Bîjagaṇita of Bhâskara* II (1150). Edited *by* Sudhakara Dvivedi (Benares) and revised by Muralidhara Jha (Benares, 1927); with the commentary, entitled *Navâṅkura*, of Kṛṣṇa (*c.* 1600) *by* D. Apte (Anandasrama Sanskrit Series, Poona, 1930). English translation *by* H. T. Colebrooke (*Algebra with Arithmetic and Mensuration from the Sanscrit of Brahmegupta and Bhascara*, London, 1817). Other commentaries in MSS: (*i*) *Bîjaprabodha by* Râmakṛṣṇa (*c.* 1648); (*ii*) *by* Sûryadâsa (born 1508). References to this treatise in our text are by pages of Muralidhara Jha's revised edition.

8. *Bîjagaṇita* of Jñânarâja (1503). MS.

9. *Bîjagaṇita* of Nârâyaṇa (1350). MS. (incomplete).

10. *Brâhma-sphuṭa-siddhânta by* Brahmagupta (628). Edited with explanatory notes by Sudhakara Dvivedi, Benares, 1902. Chapters xii and xiii of the work dealing respectively with arithmetic and algebra have been translated into English *by* H. T. Colebrooke (*Algebra with Arithmetic and Mensuration* etc.). Commentary *by* Pṛthudakasvâmî (860). MS. (incomplete).

11. *Bṛhajjâtaka* of Varâhamihira (505). Edited with the commentary of Bhaṭṭotpala (966) *by* Rasikmohan Chattopadhyaya (Calcutta, 1300, B.S.); *by* Sitaram Jha (Benares, 1923).

12. *Bṛhat-saṁhitâ by* Varâhamihira (505). Edited *by* H. Kern (Calcutta, 1865); *by* Sudhakara Dvivedi, with the commentary of Bhaṭṭotpala (Vizianagram Sanskrit Series, 2 Vols., Benares, 1895). English translation by H. Kern (See his collected works).

13. *Dhyânagrahopadeśa* of Brahmagupta (628). Edited *by* Sudhakara Dvivedi and published as an appendix to his edition of the *Brâhma-sphuṭa-siddhânta*.

14. *Gaṇita-kaumudî* of Nârâyaṇa (1356). MS.

15. *Gaṇita-mañjarî by* Gaṇeśa, son of Ḍhuṇḍhirâja (1558). MS.

16. *Gaṇita-tilaka by* Śrîpati (1039). Edited with the commentary of Siṁhatilaka Surî (*c.* 1275) *by* H. R. Kapadia (Gaikwad Sanskrit Series, Baroda, 1935).

17. *Gaṇita-sâra-saṁgraha by* Mahâvîra (850). Edited with English translation and notes *by* M. Rangacarya, Madras, 1912.

18. *Graha-lâghava* of Gaṇeśa Daivajña (*c.* 1545). Edited with the commentaries of Mallâri, Viśvanâtha and his own *by* Sudhakara Dvivedi, reprinted, Bombay, 1925.

19. *Karaṇa-kutûhala by* Bhâskara II (1150). Edited with the commentary of Sumatiharṣa *by* Madhava Sastri, Bombay, 1901.

20. *Karaṇa-paddhati by* Pathumana Somayâji (1733). MS.

21. *Kâtyâyana Śulba* Sûtra *by* Kâtyâyana (*c.* 400 B.C.). Edited with explanatory notes *by* Vidyadhar Sharma, Benares, 1928.

22. *Khaṇḍa-khâdyaka by* Brahmagupta (628). Edited with the commentary of Âmarâja (*c.* 1250) *by* Babua Misra (Calcutta, 1925). English translation with notes and comments *by* Prabodh Chandra Sengupta (Calcutta, 1934). Other commentaries in MSS: (*i*) *by* Pṛthudakasvâmî (incomplete); (*ii*) *by* Varuṇa; (*iii*) *by* Bhaṭṭotpala.

23. *Laghu-Bhâskarîya by* Bhâskara I (522). Commentary *by* Śaṅkaranârâyaṇa. MSS.

24. *Laghu-mânasa by* Mañjula (932). Commentaries *by* (*i*) Pṛthudakasvâmî (964); (*ii*) Parameśvara (1430). MSS.

25. *Lîlâvatî by* Bhâskara II (1150). Edited with notes *by* Sudhakara Dvivedi, Benares, 1910. Translations: (*i*) *by* H. T. Colebrooke (*Algebra with Arithemtic and Mensuration* etc.); (*ii*) *by* J. Taylor (*Lilawati*, Bombay, 1816). Colebrooke's translation has been re-edited with critical notes by Haran Chandra Banerji (*Lîlâvatî*, 2nd ed., Calcutta). Commentaries in MSS: (*i*) *Buddhivilâsinî by* Gaṇeśa (1545); (*ii*) *Gaṇitâmṛtasâgarî by* Gaṅgâdhara (1432); (*iii*) *Gaṇitâmṛtalaharî by* Râmakṛṣṇa (1339); (*iv*) *Manorañjana by* Râmakṛṣṇadeva; (*v*) *Gaṇitâmṛtakûpikâ by* Sûryadâsa (1541); (*vi*) *Cintâmaṇi by* Lakṣmîdâsa (1500); (*vii*) *Nisṛṣṭadûtî by* Munîśvara (1608). References to *Lîlâvatî* in our text are by the pages of Dvivedi's edition.

26. *Mânava Śulba* Sûtra *by* Manu, MS. English translation *by* N. K. Mazumdar (in *Journal of the Department of Letters in the University of Calcutta*, VIII, 1922).

27. *Mahâ-Bhâskarîya by* Bhâskara I (522). Commentary *by* (*i*) Sûryadeva (12th century); (*ii*) Parameśvara (1430). MSS.

28. *Mahâ-siddhânta by* Âryabhaṭa II (950). Edited with explanatory notes by Sudhakara Dvivedi, Benares, 1910.

29. *Pañca-siddhântikâ by* Varâhamîhira (505). Edited with a commentary in Sanskrit, translation into English and critical notes *by* G. Thibaut and Sudhakara Dvivedi, Benares, 1889.

30. *Pâṭî-sâra by* Munîśvara (born 1603). MS.

31. *Sadratnamâla by* Śaṅkaravarman. MS.

32. *Siddhânta-śekhara by* Śrîpati (1039). Edited *by* Babua Misra, Calcutta, Vol. I (1932), containing chapters i-xii of the text, with the commentary of Makkibhaṭṭa (1377) on chapters i-iv and that of the editor on the rest.

33. *Siddhânta-śiromaṇi by* Bhâskara II (1150). Edited with the author's own gloss (*Vâsanâbhâṣya*) *by* Bapu Deva Sastri, Benares); *by* Muralidhara Jha with the commentaries, *Vâsanâvârtika* of Nṛsiṃha (1621) and *Marîci* of Munîśvara (1635), Vol. I (containing chap. i of the *Gaṇitâdhyâya*) (Benares, 1917); *by* Girija Prasad Dvivedi, with original commentaries in Sanskrit and Hindi, Vols. I and II (Lucknow, 1911, 1926). English translation of the text only by Bapu Deva Sastri and Wilkinson (Calcutta, 1861).

34. *Siddhânta-tattva-viveka by* Kamalâkara (1658). Edited with the *Śeṣa-vâsanâ* of the author, *by* Sudhakara Dvivedi, Benares, 1885.

35. *Śiṣya-dhî-vṛddhida by* Lalla (598). Edited *by* Sudhakara Dvivedi, Benares, 1886. Commentary *by* Mallikârjuna Surî (1179). MS.

36. *Sûrya-siddhânta*. Edited *by* F. E. Hall and Bapu Deva Sastri with the commentary of Raṅganâtha (Calcutta, 1859); *by* Sudhakara Dvivedi with an original commentary in Sanskrit (Calcutta, 1909-11). Translation into English, with critical notes *by* E. Burgess and W. D. Whitney; into Bengali with critical notes *by* Vijñânânanda Svâmî.

37. *Tantra-saṅgraha by* Nîlakaṇṭha (1500). Commentary *by* an unknown writer. MS.

38. *Triśatikâ by* Śrîdhara (750). Edited *by* Sudhakara Dvivedi, Benares, 1899. The principal rules in this text leaving the illustrative examples and their solutions, have been translated into English *by* N. Ramanujachariar and G. R. Kaye and published with notes and comments ("Triśatikâ of Śrîdharâcârya," *Bibl. Math.*, XIII (3), 1912-13, pp. 203ff.). Dvivedi's text is apparently incomplete. The manuscript in our collection, though not perfect, contains a few more rules and examples.

INDEX

T

U